Springer Biographies

The books published in the Springer Biographies tell of the life and work of scholars, innovators, and pioneers in all fields of learning and throughout the ages. Prominent scientists and philosophers will feature, but so too will lesser known personalities whose significant contributions deserve greater recognition and whose remarkable life stories will stir and motivate readers. Authored by historians and other academic writers, the volumes describe and analyse the main achievements of their subjects in manner accessible to nonspecialists, interweaving these with salient aspects of the protagonists' personal lives. Autobiographies and memoirs also fall into the scope of the series.

More information about this series at http://www.springer.com/series/13617

Olival Freire Junior

David Bohm

A Life Dedicated to Understanding the Quantum World

 Springer

Olival Freire Junior
Salvador, Bahia, Brazil

ISSN 2365-0613   ISSN 2365-0621   (electronic)
Springer Biographies
ISBN 978-3-030-22717-3   ISBN 978-3-030-22715-9   (eBook)
https://doi.org/10.1007/978-3-030-22715-9

This Springer imprint is published by the registered company Springer Nature Switzerland AG
The registered company address is: Gewerbestrasse 11, 6330 Cham, Switzerland

Preface

I have been pondering David Bohm's life and works since my graduate studies at the University of São Paulo, in Brazil, thirty years ago. I discussed his life and works with the late Amelia Hamburger and Alberto Luiz da Rocha Barros, and Michel Paty, all who knew him, and Osvaldo Pessoa Junior. I was fascinated by the existence of his heterodox interpretation of quantum mechanics which still led to the same experimental predictions obtained by orthodox quantum theory. Furthermore, I was struck by his life as it mirrored some of the dramas of the twentieth century: early affinity with Communism, victim of McCarthyism and lasting exile from the USA, rupture with Communism and later rapprochement to the Indian thinker Jiddu Krishnamurti. In addition, in the early 1950s, Bohm lived in São Paulo for almost four years during the first stage of his peregrination as an American expatriate. Bohm's life and works have since permeated my life. However, I was never fully at ease with Bohm's persona, the manner in which many regarded Bohm's thinking. Some simply considered his work on the foundations of quantum mechanics worthless while others, mainly non-physicists, saw Bohm's work on quantum mechanics and philosophy as a panacea to be used everywhere, no matter what the subject. Indeed, Bohm is known in a wider cultural circle than that of physics itself; his admirers include several New Age counter-culture followers. Most of these admirers think Bohm considered quantum mechanics a universal solution for many issues. Nothing could be further from the truth, Bohm spent most of his life in the quest for understanding the quantum world, a quest he left unfinished. Both sides were wrong, I thought. Bohm's works contributed to enhance the research on the foundations of quantum mechanics, one of the marvelous pages of physics, even if we do not agree with some of his scientific ideas. Even Bohm himself was never comfortable with his own ideas on quantum mechanics. Till the last day of his life, he strived to understand the quantum world; thus, any farfetched extension of his ideas to any other subjects was far from his own scientific practice. This book is thus in first place my attempt to deal with these conflicting trends and to present, as a scholar work in the field of history of science, the whole life and works of this extraordinary scientist. It is an attempt to understand Bohm's role in the twentieth-century history of science and ideas.

The legacy of David Joseph Bohm (1917–1992) to physics may be stated in a nutshell: He was a physicist who made many and lasting contributions particularly in subjects such as plasma, metals, and quantum mechanics; he was one of the discoverers of the Aharonov–Bohm effect and suggested alternative interpretations of quantum mechanics. He was undoubtedly one of the major twentieth-century physicists. As for his legacy to quantum physics, rather than for one specific and lasting contribution, he should be better acknowledged for his continued effort to emphasize the relevance of the research on foundations for the future of physics. As to Bohm's legacy to quantum mechanics, a few considerations made by the physicist and historian of physics Max Jammer are valid. He finished his book *The Philosophy of Quantum Mechanics,* in 1974, stating that the interpretation debates in this physical theory were "a story without an ending," and appealed to the French moralist Joseph Joubert to conclude: "It is better to debate a question without settling it than to settle a question without debating it." Bohm's main legacy in quantum mechanics, I think, was his contribution to keep this debate alive in times when many of physicists thought it should be closed. Indeed, those physicists thought all the foundational issues had already been solved by the founding fathers of quantum theory. However, this was a mistake. Nowadays, we know about quantum entanglement, which is the basis of quantum information, thanks to the works of people like Albert Einstein, Bohm, John Bell, Abner Shimony, John Clauser, Alain Aspect, Anton Zeilinger, and others, who worked on foundations and interpretations of quantum mechanics.

Bohm shared with Einstein the hope of reconciling quantum mechanics with scientific realism. However, their proposals diverged. Bohm also attempted to recover causal laws in the quantum domain. Later he realized that the success and stability of quantum mechanics were evidence of a new kind of physical theory, which he called implicate order to contrast with the explicate order we know from classical theories in physics. Since the mid-1980s, there has been renewed interest in Bohm's ideas on quantum mechanics and this is related to the growing interest in the foundations of this physical theory. Nowadays, research on foundations is required either to develop quantum information or to connect quantum mechanics with gravity in the domain of quantum cosmology. Bohm's ideas are one of the ways to approach these domains.

Bohm's ideas were not limited to physics. His philosophical reflections ranged from the role of causality and chance in modern physics to the role of creativity and dialogue to enhance human sociability on our planet and included reflections on consciousness. However, while several of his contributions to physics are today part of the physics common ground, his philosophical reflections and some of his scientific speculations are more controversial, which is not surprising at all given that ideas in this domain of enquiry are normally more open to debate. His scientific and philosophical work was developed constrained by some of the major events of the twentieth century. Much of this work took place at the height of the cultural/political upheaval in the 1950s and 1960s, which led him to become the most notable American scientist to seek exile in the last century. The story of his life is as fascinating as his ideas on the quantum world are appealing.

Despite all of this, there is no one-volume summary of Bohm's life and works. Previous works have dealt with Bohm's life, the political context in which he lived and worked, and parts of his scientific works, but so far there has been no compelling scientific biography of David Bohm, as I will argue in Chap. 1. This was one of the reasons I chose this subject for a book when I was approached by Angela Lahee, Executive Editor at Springer, to suggest I write a biography of one the characters from my book, *The Quantum Dissidents—Rebuilding the Foundations of Quantum Mechanics 1950–1990*, published in 2015. Thus, this book is primarily intended for physicists and physics students as well as historians and philosophers of science who want to frame Bohm's ideas and life in twentieth-century history, while also intended for wider circles of his admirers. In the writing of this book, I have tried to achieve a trade-off between conceptual accuracies, while presenting Bohm's scientific ideas, with the minimum, almost nothing, of mathematical formalism. I hope the latter readers do not feel at sea in some parts of the book, and I hope they will be rewarded with a better picture of the scientist David Bohm if they persist in the reading of the book.

I was well placed to meet the challenge of writing this biography as I said I have been thinking about Bohm's work and life since my doctoral studies in history of science in the early 1990s. I chose the reception of Bohm's causal interpretation of quantum mechanics as a topic for my research. In fact, my choice had been in part determined by Bohm's own life as he had worked at the University of São Paulo, in Brazil, from the late 1951 to the early 1955. He had left the USA because he was a victim of McCarthyism, lost his position at Princeton, and was unable to obtain another position at any American university. He chose Brazil because there had been Brazilian physicists at Princeton, Jayme Tiomno, and Leite Lopes, who had invited him to work in São Paulo. In São Paulo, the American consulate confiscated his passport and told him they would hand it back only when Bohm returned to the USA, which Bohm was afraid of, due to the rising McCarthyism. In order to travel abroad Bohm applied for Brazilian citizenship, after which the USA withdrew his American citizenship. This situation lasted more than thirty years, and it was only in 1986 that he was able to recover it. As a result, Bohm lived as a Brazilian citizen and an American expatriate for thirty years.

In January 1992, I had the opportunity to talk to Bohm personally, in London. He passed away months later, and we did not have another meeting. Our conversation was not very fruitful for my doctoral research as everything he told to me had already been published elsewhere. I was at the beginning of my Ph.D. and did not yet have full command of the oral history research techniques. The interviewer needs to know more about the subject than the interviewee himself in order to obtain new information. From this meeting, the only material reminiscence is a few words and his signature in one of my books (by him, of course).

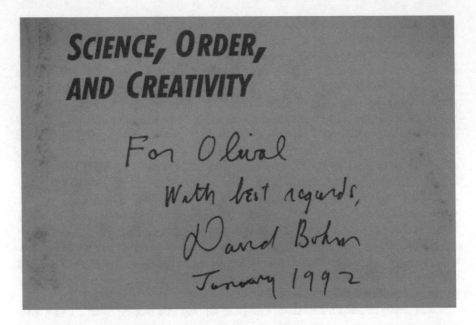

SCIENCE, ORDER, AND CREATIVITY

For Olival
With best regards,
David Bohm
January 1992

However, I remember clearly the first question he asked me. It was about the situation of the poor and hungry children who were living on the streets without shelter nor schooling, in Brazil. He was referring to this evidence of Brazilian huge social inequalities, the street kids, or "meninos de rua" in Portuguese. In hindsight, it was clear that Bohm's social consciousness was still acute. It had been forged in his youth in Wilkes-Barre, PA, in the USA where child labor in the coal mines had been the norm till the first decade of the twentieth century and unemployment was rampant in the years of the Great Depression.

Salvador, Brazil   Olival Freire Junior

Contents

Chapter 1
Introduction: Living Through Cold War Storms, Attempting to Understand the Quantum

> When I went to work with J. Robert Oppenheimer, I found a more congenial spirit in his group. For example, I was introduced to the work of Niels Bohr and this stimulated my interest, especially in the whole question of the oneness of the observer and the observed. [...] I can still recall the many discussions I had on matters like this which had the effect of setting me on the course I'm still following today.
> David Bohm, in Bohm and Peat, *Science, Order and Creativity*, 1987

Biographies are usually of interest because we may learn from them about the life of a person, their singularities in the plethora of humankind, as well as about the times in which the person lived. This is the case of David Joseph Bohm, a physicist who was born in the US, in 1917, educated there, lived in Brazil, Israel and the UK where he passed away in 1992. He was an outstanding theoretical physicist with achievements in standard science and breakthroughs in the way we obtain predictions and interpret the most successful physical theory, the quantum theory. His life was entangled with two major intellectual trends which prevailed in the twentieth century, Communism and Eastern traditions. In fact, he was first associated with the US Communist Party, thus Soviet-inspired Communism, and later with the Indian thinker Jiddu Krishnamurti. However, he faced obstacles both in the scientific community and in the political arena. Indeed, his proposals to an alternative interpretation of quantum mechanics were initially poorly received among his fellow physicists, which only changed when the overall physics community changed its appreciation towards the research on the foundations of quantum mechanics, from the 1970s on. Commitment to Communism and works close to the American atomic project at Berkeley, during WWII, cost Bohm his professional stability and permanence in the US at the apex of McCarthyism. He left the US and lived as an expatriate in Brazil, then Israel, and later in the UK. In the meantime, reflecting the Cold War times, the US government withdrew his American citizenship, which he only recovered in the mid-1980s. Proximity to mystical thinkers and Krishnamurti came as a follow-up of his break with Communism, after the 1956 Soviet invasion of Hungary. In addition, his philosophical and scientific views were never fully independent of the contexts

© Springer Nature Switzerland AG 2019
O. Freire Junior, *David Bohm*, Springer Biographies,
https://doi.org/10.1007/978-3-030-22715-9_1

where he lived. Through Bohm's biography we may thus learn about twentieth century history, physics, and the history of physics. We shall also meet the person and his ongoing anxieties, both with the understanding of the quantum and the attempt to frame it in wider cultural frameworks. Biographies are like windows, or points of perspective, through which a life and its times can be seen.

In this introductory chapter to David Bohm's biography we are going to see first an overview of his ideas and life and then methodological considerations and sources which were helpful during the writing of this book. They concern biographies as a genre in history, approaches in history such as the history of ideas and cultural history, and the sources for studies on David Bohm. From Chaps. 2–7, the biography itself is presented in chronological order, the seventh chapter being an epilogue about how Bohm's ideas have been appropriated and developed since his death. The final chapter, the eighth, is an attempt to use a quantitative criterion to assess the influence of Bohm's works among physicists. This is done through a scientometric essay about the number of papers citing Bohm's own papers.

1.1   An Overview of Bohm's Life and Works

Bohm worked almost entirely dedicated to the foundations of quantum mechanics, particularly on the interpretation of this theory, for more than 40 years. His most original and heterodox contribution to quantum physics may have been the elaboration of the causal interpretation of quantum physics, published in 1952, which departed significantly from the standard theory in its conceptual and philosophical assumptions but still arrived at the same predictions, at least in the non-relativistic domain. Departure from the standard theory came as Bohm assumed a realist point of view about the meaning of quantum mechanics and recovered the determinism for quantum phenomena, which had been discarded by the standard interpretation of this theory. The standard interpretation was also known as the complementarity point of view and it was the Danish physicist Niels Bohr who developed it, but there were important different views among the physicists who supported it. With his interpretation Bohm opened the way for alternative interpretations of quantum mechanics and showed that quantum mechanics could be supplemented by additional variables. This way he disproved by this counter example the proof against the possibility of additional variables in quantum mechanics which had been proposed by the mathematician John von Neumann in the early 1930s. Furthermore, and more importantly, Bohm's interpretation inspired John Bell to develop, a decade later, what is now called Bell's theorem. This theorem, and the series of its experiments, which are being performed even nowadays, have led the physicists to accept entanglement, that is, the quantum correlation among systems which are far away one from the other, as a physical effect, that is predicted from the theory and confirmed by experiments. Entanglement has become the physical effect on the basis of quantum information, a field which has blossomed since the early 1990s and promises the use of quantum mechanics to obtain more powerful computers and cryptography in years to come.[1]

However, Bohm's proposal was poorly received at the time. Research in foundations without new predictions was then considered more philosophy than physics. Treating foundations as just philosophical issues enhanced professional bias in the physics community against the subject, which was thus considered the realm of philosophical research and not research in physics. Furthermore, the common view among physicists in the 1950s was that foundational issues in quantum mechanics had already been solved by the creators of this physical theory. Thus there were no unsolved questions that deserved the energy of physicists, particularly young talents. In fact, the subject was considered, in the best case, a marginal topic on the physics research agenda. I have used the term quantum dissidents to designate the physicists who through their work pushed this subject from the margins to the mainstream of physics. Paramount among these quantum dissidents was David Bohm, the subject of our biography. In addition, Bohm hoped to obtain a relativistic generalization of his causal interpretation, however, he failed to do so, at least in the first years after the publication of his alternative interpretation. All this together may have alienated him from the kind of physics being developed in the 1950s, and led him to a certain isolation among his fellow physicists.[2]

Despite this isolation, Bohm was able to make another contribution to our understanding of quantum mechanics, which was later recognized as a major breakthrough. Indeed, in the late 1950s, he and his doctoral student Yakir Aharonov published their seminal paper on the understanding of the role of phases and electromagnetic potentials in the quantum description of the world. They suggested the so far unknown Aharonov-Bohm effect, meaning that quantum mechanics may lead a magnetic field to act on a charge even in regions where this field is null. The suggestion of this effect stirred up a number of theoretical and experimental works and it is now well grounded on the physicists' toolkit of quantum phenomena. Much of the recognition of Aharonov as a first-ranking physicist came from the wide acceptance of this work.[3]

Later on, in the early 1960s, Bohm abandoned the causal interpretation and moved to the program he named wholeness and order, which was fleshed out, through his collaboration with Basil Hiley, Bohm's collaborator at Birkbeck in London, as a highly mathematical approach. Working with Hiley, he looked for the most basic algebraic structures from where space, time, and quantum theory might emerge. In the early 1980s Bohm saw his initial interpretation revived by some of his students who were able to produce computer-created graphs of trajectories and potentials obtained from the causal interpretation of quantum physics. He spent his last years trying to reconcile these different approaches to the quantum.[4]

---

[1] Bohm (1952).

[2] Freire Junior (2015a).

[3] Aharonov and Bohm (1959).

[4] Philippidis, Dewdney et al. (1979); Bohm and Hiley (1993).

Bohm lived his life attempting to understand the puzzles of quantum mechanics. Since the late 1940s, when he began to systematize his views to write the textbook *Quantum Theory* (Bohm 1951) till his last days with Basil Hiley in the writing of the book *The Undivided Universe* (Bohm and Hiley 1993), there were more than 40 years dedicated to this intellectual challenge. In the history of twentieth century physics, the time and energy he spent on this may be compared to that of Einstein, Bohr, and Schrödinger, among other great physicists. As a younger physicist, he was less prepared to face the isolation derived from the manner most physicists received his alternative interpretation for quantum mechanics. Bohm's endurance, however, was formidable. To think about the meaning of the quantum and its implications for science was his bread and butter for decades. This endurance may illustrate the double motivation I had to adopt the term quantum dissidents. The first motivation considered that physicists such as Bohm, Bell, and others, were critical of what they perceived as the complementarity interpretation, or the usual interpretation of quantum mechanics. However, they did not share a unique alternative interpretation of quantum mechanics. They shared the professional and intellectual attitude that issues in foundations of quantum mechanics were worthy of pursuit as part of a professional career in physics, and that denying this was a dogmatic attitude. The second motivation, however, came from the analogy with major political dissidents in the twentieth century. When I first thought of this term I had in mind the examples of Nelson Mandela, Luís Inácio Lula da Silva, and Martin Luther King. They, or the cause they embraced won, at least in the medium term. This was possible, among other factors, thanks to the moral endurance of the dissidents.[5] In the case of quantum mechanics, the moral value of the resilience is similar. The foundations of quantum mechanics eventually became a respectable field of research and it has contributed to a better understanding of the quantum. This was possible, among other reasons, owing to the moral, intellectual, and professional endurance of the quantum dissidents. David Bohm was a paramount character among these dissidents sharing their common features, including their perseverance.

Bohm's contributions to physics were not limited to foundations of quantum mechanics. He was a doctoral student of Julius Robert Oppenheimer at Berkeley in the early 1940s, and worked on subjects which were of interest to the Manhattan Project, the American atomic project. He emerged from the war effort considered a highly promising American theoretical physicist and was then hired by Princeton University being recommended by the American physicist John Archibald Wheeler. While working on subjects which were part of the war effort, he dealt with the problem of electrical currents passing through a gas in magnetic fields, which led him to the study of plasmas. When the war ended, Bohm and his graduate students Eugene Gross and David Pines worked on plasmas, developing the approach called "collective variables" and applied it to the study of current in metals, elaborating a quantum approach to the phenomenon using the same collective variable resource he had successfully applied in the classical treatment of plasmas. Later, this approach

---

[5] I coined the term quantum dissidents in 2009 (Freire Junior 2009). Ten years later, regressive trends in many parts of the world and market pressures on researchers suggest taking the "medium term" with a grain of salt as progressive accomplishments are never assured forever.

was used in superconductivity by Pines, and in nuclear physics by Aage Bohr and Ben Mottelson. The jointly-authored papers of Bohm, Pine, and Gross became landmarks in the field also known as many-body physics.

Bohm's achievements in science ranked him as a very influential physicist in the twentieth century. He was elected Fellow of the Royal Society, had a *Festchrift* with papers signed by four Nobel prize winners, and had his work on the reinterpretation of quantum mechanics featured in the *Physical Review* centennial anniversary volume. Using the metrics of the number of citations, the extent of his influence can also be traced. At the end of 2018, ten of his papers had obtained more than 300 citations, one of them, on the Aharonov-Bohm effect had over 4300 citations, and another, on the causal interpretation, had almost 3000 citations. These are impressive numbers, by any standard, for twentieth century physics papers. He also wrote a number of well received books, among them a textbook on quantum mechanics still considered today a good textbook on the matter. However, his reputation fluctuated. From being considered a promising young physicist, his reputation was damaged by the adverse reception of his work on the foundations of quantum mechanics; exemplifying the bias against research on foundations in the 1950s. After the experiments related to Bell's theorem, interest in his early work was slowly resumed. Ultimately, this theorem and its experiments set entanglement as a strictly quantum feature and shed light on the equivalence between standard quantum mechanics and Bohm's interpretation. Both of them feature quantum entanglement as one of their features while they provide different explanations for it. Thus, due recognition for his outstanding contributions came later, as noted by his old friend the physicist Melba Phillips, in the early 1990s: "It is too bad, very sad indeed, that he did not live to see how his reputation has shot up recently. His interpretation of quantum mechanics is becoming respected not only by philosophers of science but also by 'straight' physicists".[6]

In hindsight how can we assess Bohm's legacy to quantum physics? In addition to his specific and lasting contributions, I think he should be acknowledged for his attitude which highlighted the relevance of the research on the foundations of this theory. John Bell's recollections about how he was driven to these issues encapsulate this feeling: "In 1952 I saw the impossible done," and "Bohm's 1952 papers on quantum mechanics were for me a revelation," referring to the appearance of the causal interpretation which was considered by current wisdom an impossible feat.[7]

Bohm developed his own scientific style from his first years of graduate studies at Caltech and Berkeley. To follow the terms used by historian Suman Seth (2010, p. 2), between a "physics of problems," illustrated by Arnold Sommerfeld, and a "physics of principles," whose paragons in the early twentieth century were Albert Einstein and Max Planck, Bohm tailored his own and distinctive style. He was more inclined towards a scientific style based on concepts and philosophical views as points of departure, freely using pictures as supports for his physical thinking, and only then going to the mathematical machinery. He was well trained in

---

[6]The Festschrift is (Bohm et al. 1987) and the textbook is Bohm (1951). Phillips letter is quoted in (Freire Junior 2015a, p. 63).
[7]Bell (1982 and 1987).

mathematics, thus mathematics coming at the end was rather a scientific and epistemological choice than a handicap. His scientific style was closer to Einstein's than to
Sommerfeld's. Furthermore, in the early 1950s, his quest for a new interpretation for
quantum mechanics was associated to a disdain for the machinery of mathematics—
renormalization techniques—built by Richard Feynman and others to deal with the
infinities plaguing quantum field theories. Later, in the 1960s, in discussions with
Jeffrey Bub, one of his former graduate students, it became explicit that Bohm did not
keep in high esteem axiomatic approaches such as those adopted by von Neumann.

From when he was a teenager, David Bohm developed a keen social consciousness.
He grew up in Wilkes-Barre, a coal mining town in Pennsylvania that was devastated
in the decade of 1930 in the wake of changes in the use of coal and the Great
Depression. Unemployment, social unrest, and labor organizations were part of the
ambiance where Bohm came of age. Later, under the influence of threats of the rising
Nazi power and the US entrance into the war, at Berkeley Bohm joined the Communist
Party. After the Second World War, in the inception of the Cold War, he was caught
up in American anxieties with Communism and the suspicion of leaks of atomic
secrets. He was called to testify before the House Un-American Activities Committee
(HUAC) and exercised his constitutional right to remain silent in order to prevent self-
incrimination and name names of colleagues. His attitude was in stark contrast with
the HUAC's expectations. He was condemned of contempt to the Congress, jailed,
released on bail and later acquitted. Despite being acquitted, Princeton University
did not renew his contract. In 1951, he left the US for a job in Brazil and later went
to Israel and eventually the United Kingdom in 1957. Reflecting US policies in the
Cold War times, his passport was apprehended by American officials in Brazil. In
order to get a passport to travel abroad, he applied and got Brazilian citizenship
but the US then cancelled his American citizenship. He only recovered it 30 years
later, in the twilight of the Cold War and through a legal procedure. He lived for
almost 3 decades as one of the most notable American expatriate scientists. In the
late 1950s, following Khrushchev's report on Stalin's crimes and the Soviet invasion
of Hungary, he broke his ideological ties with Marxism. Following this rupture, he
moved towards a rapprochement with mystical thinkers such as George Gurdjieff,
born in Armenia, then part of Russian empire, and the Russian Peter Ouspensky.
Furthermore, Bohm was strongly drawn to the writings of Jiddu Krishnamurti. This
philosopher was born in India and educated by the esoteric movement Theosophy to
be its new world religious leader. Later Krishnamurti rejected this connection and
went his own way of expressing his teachings. From that time on Bohm moved away
from his political interests and focused on self-improvement, particularly following
Krishnamurti's lessons and dialoguing with him. He then became an iconic figure in
the New Age culture of the 1970s. His life illustrates thus much of the political and
cultural turmoil of the times as it reflected the circumstances and vicissitudes of the
twentieth century, in politics and culture at large. In addition, these commitments
as well as their changes exemplify his anxieties with values and wider intellectual
frameworks.

Bohm's life circumstances and work in physics were intertwined enough to
lead commentators to highlight the strong influences of the former on the latter.

Alex Kojevnikov suggested his approach to plasma and electrons in metals in terms of collective variables reflected his early Marxist commitments. Christian Forstner saw his move towards the causal interpretation as conditioned both by his Marxist views and his isolation from the Princeton community during his persecution under McCarthyism. I suggested his moving away from the causal interpretation, in the late 1950s, was constrained by his break with the ideological ties of Communism. The connections between his wholeness and implicate order approach, which he adopted from the 1970s on, and the thoughts of Krishnamurti are conspicuous. Bohm himself saw it this way and wrote on it, for instance in his 1980 book *Wholeness and Implicate Order*. The complexities of these connections will be discussed at greater length in Chaps. 5 and 6. In an analogous manner, his return to the causal interpretation was strongly influenced by John Bell, either through his theorem and its implications or by Bell's support for Bohm's hidden variables interpretation, as well as by Hiley's students, who produced computer-based pictures based on the causal interpretation. Not all of the connections are equally grounded on evidence and some lack either documentary evidence or plausibility or both. This was the case of the suggestion that the poor reception of the causal interpretation was influenced by his persecution during McCarthyism; a suggestion I found unfounded.[8] Thus, on the one hand, the study of Bohm's scientific work requires tracing the contexts in which he lived as the former was constrained by the latter and the latter was also shaped by the impact of the former. On the other hand, Bohm was above all a physicist, thus, the study of the contexts without due attention to his scientific work would dilute what he saw as his main contributions.

1.2 Biographies, History, and History of Science: Methodological Notes

A biography of a great scientist is a professional endeavor which brings together different areas of history. Biography, history, history of science, history of ideas, and cultural history were the fields I had to walk through during the research for and the writing of this book. Let me make a few short comments about these subjects.

While biographies have been bestsellers for decades, they are latecomers to historiographical scholarship. The reason for this has been professional distrust towards a kind of historical work usually related to the description of heroes and saints. Interestingly, it is from here that the term hagiography comes, the deadliest of sins committed by professional historians. The history of science is no different, many biographies of scientists have been published but the genre has been suspect for history of science scholarship. Replacing heroes and saints with individual geniuses, most biographies in science were idealized descriptions of these outstanding scientists. As noted by the historian of science Thomas Söderqvist, influences from philosophy and later from sociology in history of science did not value biography as a genre. The same may be

---

[8]Kojevnikov (2002), Forstner (2008), Freire Junior (2015a).

said of the post-structuralist trends with its deconstruction of the subject. Evidence of this distrust is that, who was writing a biography of a scientist, could ask, twenty years ago: "What is the legitimate place of biography in history of science? Is it simply a sort of sophisticated entertainment, the scientist's bedside companion after the daily torments in the laboratory or at the desk, and thus better handed over to novelists, or is biography a possible and valuable scholarly pursuit in itself?"[9]

It was more recently that the genre was fully recovered through the approach the French historian François Dosse called the hermeneutical stage in the historiographical fortune of biographies. The historian has no ambition to represent the individual as a coherent character in his times, instead, fractures, contradictions, and conflicts should be highlighted. In scholarship, however, new trends cannot be only rhetorical, they require good examples. During the research for this book I had in mind two biographies which dealt with the current standards of the history as a discipline particularly well. The first was *Never at Rest*, Newton's biography by Richard Westfall. While most of Newton's biographies portray the genius, Westfall's book illustrates how off scale individual and contributions can be accommodated in a biographical portrait full of other human features. The second was Jacques Le Goff's portrait of Saint Louis, King of France and later a Catholic saint. I had Le Goff's work in mind not so much because of its content, a thirteenth century king and saint is a too distant comparison for a twentieth century physicist, but because the late dean of French historians presented a number of methodological concerns which may be useful for any biography as a genre in history. Initially Le Goff discussed the ongoing debate among professional historians about the return of the narrative in the historiographical scholarship, many of them condemning this return. According him, "all history is narrative because, placing itself in time by definition, in succession, it is necessarily associated with narration. But that is not all. First, contrary to what many—even many historians—believe, there is nothing immediate about the narrative. It is the result of an entire series of intellectual and scientific operations that one has every reason to expose, in other words, to justify". On the risks of the biographical genre, Le Goff was well aware of them, stating, "[The biographical narrative] also induces an interpretation and represents a serious danger. Jean-Claude Passeron has pointed out the risk of the 'excess of meaning and coherence inherent in any biographical approach". Furthermore, still following Le Goff, "what he [Passeron] calls the 'biographical utopia' not only consists in the risk of believing that 'nothing is meaningless' in biographical narrative without selection and criticism, but perhaps even more in the illusion that it authentically reconstitutes someone's destiny". As Le Goff's readers were familiar with Pierre Bourdieu's critique of what he called biographical illusion, Le Goff acknowledged the threat and explained how he dealt with it: "I have tried several times to escape the constraining logic of this 'biographical illusion' denounced by Pierre Bourdieu. Saint Louis did not ineluctably proceed toward his destiny as a saintly king in the conditions of the thirteenth century and in following the dominant models of his time. He formed himself and formed his era as much as he was formed by it. This construction was made up of chance and hesitation

---

[9]On biographies in the history of Science, see Kragh (1987). Söderqvist (1996, 46).

over different choices". The readers of my David Bohm biography will note that I did not ignore his hesitations, his anxieties, his frustrations with physics and politics at large, and his toing and froing about how to interpret quantum mechanics. This approach may not please some members of the circle of Bohm's admirers, however, I think otherwise. As in other fields of human experience, it is better to deal with the "real" portraits than with idealized representations.[10]

Still following Le Goff's reflections, he made two considerations I took into account: "Finally, as [Jorge Luis] Borges stated, a man is never really dead until the last man who knew him is dead in turn, [...]. The biography I have written therefore continues up to Saint Louis' definitive death, and no further". Still, "I have dedicated the second part of this work to the critical study of the production of the memory of the saintly king by his contemporaries". Unable to strictly follow the recommendation for a personage who lived mostly in the second half of the twentieth century, I attenuated the problem dealing with Bohm's disciples in Chap. 7, the Epilogue. This was also the reason I used, when possible, oral histories as sources of my research. Finally, in Chap. 8, I tried to see how Bohm was seen by his contemporaries through the use of the lens of scientometry, drawing from the number of citations of Bohm's papers by his contemporaries.[11]

These considerations led me to the first type of questions I had to deal with. How did contextual constraints affect Bohm's personal and intellectual choices? Furthermore, what kind of choice did he have at each crucial moment he faced and how did he act in response? These questions will haunt us throughout this book. A biography of a scientist who worked at the cutting edge of his discipline requires an account of the content of the science produced by this scientist and the changes he was able to instigate. Except for his early works, on plasmas and the approach of collective variables, almost all of Bohm's scientific productions were dedicated to understanding the foundations and the possibly different interpretations of this physical theory. As we know, the interpretation of this theory, focused as it was on the understanding of light, microscopic matter, and their interactions, occupied most of Niels Bohr and Albert Einstein lives, while they adopted opposite stances on many of these issues. Figure 1.1, a sculpture of Bohr and Einstein in a park in Moscow, may thus illustrate the backdrop of Bohm's work. It may give us an idea of the scale of the problems Bohm himself chose to tackle in his professional career.

As Bohm chose this subject, the foundations of quantum theory, for his research, I had to take into account the way in which this subject dramatically changed in content and prestige from the 1950s to the 1990s. My task was made easier as this was the subject of my previous book entitled *The Quantum Dissidents—Rebuilding the Foundations of Quantum Mechanics 1950–1990* and I had Paul Forman's and Max Jammer's seminal works on the long controversy over the interpretation of quantum mechanics and the vast literature on the history of quantum mechanics as useful resources. In the *Quantum Dissidents* I valued the manner in which the

---

[10]Dosse (2010). Westfall (1980). Le Goff's citations are in the introduction of Le Goff (2009).

[11]All citations of Le Goff are in the introduction of his Saint Louis' biography (Le Goff, 2009), particularly on pages xxv, xxxi–xxxii.

Fig. 1.1 Bohr and Einstein's debate, the major motivation for Bohm's work, portrayed in the artistic imagination—Sculpture in Park Muzeon, Moscow. *Credits* Photo by Climério P. da Silva Neto

individuals (who were the protagonists of these changes) indeed acted. The changes and their contexts, described in the book, were also influential in Bohm's case, and for this reason a full chapter of the book was dedicated to Bohm.[12] As the changes in the appreciation of foundations of quantum mechanics among physicists constrained what happened with Bohm's varying scientific reputation, in this biography I had to exploit the singularities of his case. Thus I had to deal with the internal consistency, but also the changes, including the ruptures, and the ambiguities in the way Bohm interpreted the quantum.

In addition to Bohm's science, particularly foundations of quantum mechanics, I had to deal with the historical contexts in which he lived, in particular the contexts which strongly shaped his own life. Thus it was a deep dive into twentieth century history, namely the years 1930s, Word War II, Cold War and McCarthyism, the rise, crisis, and decline of the influence of Communism throughout the world, and the appeal of eastern thinkers such as Krishnamurti in western societies.[13] I also mobilized resources from what historians Suman Seth and Massimiliano Badino have called the New Intellectual History, strongly influenced by cultural history. The last resource I have used throughout my research was the transnational approach to the history of science.[14] This was necessary due to the very difficulty of labeling David Bohm: was he an American physicist? Yes and no. He was trained and blossomed as a physicist in the US but he lived most of his active scientific life in exile or as an expatriate, in Brazil, Israel, and the UK. Thus following Bohm while he crossed national borders, and paying attention to the obstacles he had to overcome to cross borders it made sense to me to use this historical approach. All these contexts in science, this long lasting quantum controversy, and in society at large coalesced in a single life, that of David Bohm, with his ambitions, expectations, hesitations, changes, achievements and frustrations. Thus, if I had to single out my major expectation while writing this biography, I would say it is similar to that of Söderqvist. He appealed to the philosopher Richard Rorty, to say what he expected from biographies in the history of science: namely they should "help scientists and non-scientists alike to strengthen their abilities to live fuller and more authentic intellectual lives". Thus, I finish this section citing Söderqvist (1996, 47–75) in extenso:

---

[12]Freire Junior (2015a). A short presentation of the book's argument is Freire Junior (2015b). Forman (1971) and Jammer (1974).

[13]It would be beyond the scope of this chapter to comprehensively identify the literature I looked to for support to understand those contexts. A summary list includes: on the Cold War and twentieth century history, Gaddis (2005), Westad (2017), Hobsbawm (1995 and 2002), and Judt (2005); on McCarthyism, Schrecker (1986 and 2002) and Wang (1999); on Marxism, (Hobsbawm 1991 and 2011), Ory and Sirinelli (2004), and Caute (1967); on Krishnamurti and Bohm, (Moody 2017).

[14]On the intellectual history and the history of physics, see Seth (2011) and Badino (2016). See also Staley (2013) and the editorial, written by Peter Galison and Andrew Warwick, to the special issue of *Studies in History and Philosophy of Modern Physics* [29(3), 1998], titled "Cultures of Theory". On the transnational approach to the history of science, see John Krige (2019). In this last book our contribution (Freire Junior and Silva 2019) exploits the case of David Bohm concerning his move from the US to Brazil.

The aim of biography is not primarily to be an aid for the history of science, nor to be a generator of case studies. Instead of adding to the 'hermeneutics of suspicion' that governs so much of today's history and sociology of science, the main purpose of science biography is, I suggest, as a genre that can provide a variety of exemplars of existential projects of individual scientists—narratives through which we can identify ourselves with others who have been confronted with existential choices and struggled with the existential conditions for living in and with science. Such life stories not only provide us with opportunities to understand ourselves, intellectually as well as emotionally, but may also change and create ourselves. Hence biographies of scientists are 'edifying'—they can help us reorient our familiar ways of thinking about our lives in unfamiliar terms, and 'take us out of our old selves by the power of strangeness, to aid us in becoming new beings.'

1.2.1 Sources for the Study of Bohm's Life and Works

The writing of this biography was eased by the vast amount of studies and resources on Bohm's life and works made available in the last four decades. They include his papers, which are catalogued and deposited at Birkbeck College, London; several interviews, most of them transcribed and deposited at the American Institute of Physics, College Park, MD, in the US; a handful scholarly and popular works, beginning in 1977; the full opus of Bohm's published articles and books; and some websites, mostly related to his relationship with Jiddu Krishnamurti. I initially address Bohm's first biography, and then move to comment on the other works and sources.

As there already is a biography of Bohm, written by F. David Peat, the first question to be raised is what is the point of writing a second biography? A trivial answer would be that many great physicists have more than one biography written on them and Bohm was a great scientist. A more substantive answer requires considering the strong and weak aspects of this biography. While well written, and extensively based on letters and interviews in addition to personal acquaintance with Bohm, this biography is more a biography of Bohm's life than of Bohm's ideas. To be more precise, in key points in the development of Bohm's quest to understand quantum theory and its philosophical implications, Peat's book is too shallow and sometimes mistaken. Let me illustrate this with a conspicuous case. Bell's theorem and its impact on physics received a mere 2-page text in the full book. Peat did not acknowledge the distinction among the two papers by Bell, which were published in the inverse order of their production, one dealing with von Neumann's proof and Bohm's causal interpretation, and the second suggesting what we call now Bell's theorem. Therefore the conundrum involved in that proof and Bohm's and Bell's works is simply ignored. In addition, the book did not discuss how Bohm reacted to Bell's theorem at all. Bohm evolved from an initial misunderstanding of Bell's theorem to a full comprehension of its implication. Furthermore, Bohm grasped these implications at a moment when there were still conflicting experimental results, those from Clauser and Freedman at Berkeley, confirming quantum mechanics and those from Holt and Pipkin at Harvard, confirming local theories. Only after Clauser's replication of Holt's experiment and mainly after Fry and Thompson experiment in 1976, did the balance incline towards quantum mechanics and its non-locality. Instead of analyzing this rich moment, Peat simply and briefly stated the following: "Soon after Bell's theorem was published, a number of experimental tests, each one more refined and each one designed to overcome possible objections, confirmed the essential nonlocality of the quantum world".[15] Thus, Peat's book ignored both the scientific content and the human drama of the ten years between the appearance of Bell's theorem and the full vindication of quantum mechanics nonlocality, which later was dubbed entanglement. In fact, characters such as John Clauser, Abner Shimony, Ed Fry, Alain Aspect, and Bernard d'Espagnat are meaningfully absent in this biography. Thus a new biography of Bohm is in order, one integrating the history of his ideas and the history of his life, all articulated in their relevant contexts.

Similar comments on the strengths and weakness of Peat's book were made by the contemporary reviewers of the book. Thus, the physicist Sheldon Goldstein praised how Peat presented the manner in which Bohm's ideas were treated by fellow physicists but remarked that "Peat's treatment of the relevant physics is not always entirely accurate," and then illustrated his point. The historian Alexei Kojevnikov, after positively commenting on how Peat inserted Bohm in the contexts of the times, concluded stating "As for Bohm's science, that awaits another, more thorough analysis". Finally, the philosopher James Cushing went along the same lines, stating, "Peat does not always represent well Bohm's scientific work itself (such as his 1952 papers on quantum theory) or its subsequent impact," and "in summary, this book does make a *prima facie* case for Bohm as a 'fascinating and important scientist' and is certainly well worth reading. But it probably has not 'given David Bohm his due.'" During the research for this book, I also realized that Peat's biography underestimated certain aspects of Bohm's personal life, for instance, the role of the high school in his coming of age and the roots of his early social consciousness.[16]

Studies in history, sociology, and philosophy of science on Bohm's life and works began to pop up while Bohm was still alive. The new sociology of science was the first to approach the subject. Trevor Pinch used the theoretical framework proposed by Pierre Bourdieu to analyze Bohm's challenge to von Neumann's proof against the existence of hidden variables in quantum mechanics. The philosopher James Cushing analyzed the position of Bohm's causal interpretation in the physics community as constrained by historical contingency. Cushing appealed both to Paul Forman's study on the inception of the acausal quantum mechanics in Weimar's Germany and the Duhem-Quine thesis on the underdetermination of scientific theories by the empirical data. He argued that Bohm's proposal became the minority interpretation only because it appeared later, in the early 1950s. The acausal standard quantum mechanics, had appeared and been accepted earlier, in the mid-1920s. Bohm's troubles with the American McCarthyism was studied by Russ Olwell and Shawn Mullet. Bohm's political commitments shaping and constraining Bohm's studies on plasma

---

[15]Peat (1997, pp. 168–170). Bell (1964 and 1966). Bohm and Hiley (1975). Peat (1997, p. 170). For a detailed discussion of the appearance of Bell's theorem and its early experiments, see (Freire Junior, 2015a), in particular Chap. 7.

[16]Goldstein (1997). Kojevnikov (1998). Cushing (1997).

and many-body systems was investigated by Alexei Kojevnikov. My own studies have been dedicated to Bohm's stay in Brazil, the early reception of his causal interpretation, and his later change of mind regarding the interpretation of quantum mechanics. Connections between Marxism and Bohm's causal interpretation was considered by Andrew Cross while Christian Forstner studied contextual influences in the inception of the causal interpretation. The philosopher Paavo Pylkkänen has investigated Bohm's philosophical approach concerning the idea of order and exploited its implications for the philosophy of mind and has edited part of Bohm's correspondence with the artist Charles Biederman. More recently, Boris Kožnjak revised Bohm's participation at the 1957 Bristol conference and Chris Talbot brought a magnificent resource to Bohm's studies, the transcription and critical edition, plus an introductory analysis, of Bohm's correspondence, in the 1950s, with Melba Phillips, Hanna Loewy, and Miriam Yevick. This list is more exhaustive in the English language but it is far from comprehensive in a wider language spectrum. For instance, in Portuguese, Rodrigo Carvalho dedicated his doctoral dissertation to Bohm's philosophical views and Rodolfo Petrônio wrote an essay on and translated *Causality and Chance*. I was lucky enough to undertake this biography at a moment when I could count on these scholar works; they were both useful for the writing of this book and sources of pleasure during their readings.[17]

The David Bohm Papers, deposited at Birkbeck College, London, was a rich resource for my research. I was able to consult them a number of times always counting on the tireless support of librarians Sue Godsell and Emma Illingworth. Bohm, however, was a poor preserver of his own correspondence. Thus I had to supplement the research at Birkbeck with a search for documents from and to Bohm in other archives. Particularly useful in this sense was the Léon Rosenfeld Papers, deposited at the Niels Bohr Archive, in Copenhagen. Throughout the book the reader will note that from time to time I cite Bohm's correspondence deposited elsewhere as I also found relevant correspondence and documents on Bohm at these other archives, for instance the Guido Beck Papers at the Centro Brasileiro de Pesquisas Físicas, Rio de Janeiro; the Costas Papaliolos Papers, at Harvard Archives; the Norbert Wiener Papers, at the MIT Archives; and the John Wheeler Papers, at the American Philosophical Society, in Philadelphia. I also could use documents concerning Bohm at Technion, in Israel, at the Universidade de São Paulo, in Brazil, at the Krishnamurti Foundation, in Ojai, CA. in the USA; and at the GAR Memorial Junior and High School, in Wilkes-Barre, PA, in the USA.

Bohm intended to write an autobiography, which never materialized. In order to do this, he was extensively interviewed by Maurice Wilkins. The results were impressive: twelve sessions of recording, 612 pages of transcripts, between 6 June 1986 and 16 April 1987, covering Bohm's entire life. That Maurice Wilkins, the "third man" in the Nobel Prize for the discovery of the DNA structure, had spent so much time interviewing Bohm is a sign of the high esteem he had for Bohm. They

---

[17]Pinch (1977). Cushing (1994). Olwell (1999). Mullet (2008a and b). Kojevnikov (2002). Freire Junior (1999, 2005 and 2011). Cross (1991). Forstner (2008). Pylkkänen (2007) and Bohm et al. (1999). Kožnjak (2018). Talbot (2017). Carvalho (2015), Petrônio (2013).

met each other at Berkeley, during World War II, when they began a solid and lasting friendship. According to Wilkins, Bohm "became a long-standing friend in England. We three [Wilkins, Bohm, and Eric Burhop] enjoyed a holiday together [...] on Lake Tahoe, and Dave and I went on to Lake Pyramid to climb the famous pyramid-shaped rock there. Eric and Dave combined their enthusiasm for scientific progress with a sensitivity and concern for the problems of human life more generally". It is noticeable that in Wilkins Papers, 23 folders are dedicated to publications concerning David Bohm.[18] Bohm interviewed by Wilkins is a mandatory source for studies on Bohm's views. This interview and a certain number of other interviews with Bohm are available for consultation at the American Institute of Physics (AIP), College Park, MD.

The collection of interviews with physicists available at the AIP is impressive and I used a number of them to collect relevant information for Bohm's biography. Furthermore, I was able to interview people, who were kind enough to save time for these interviews, and deposit them at the AIP. They were Basil Hiley, Chris Dewdney, Franco Selleri, Alain Aspect, Anton Zeilinger, Nicolas Gisin, Yanhua Shih, and Sheldon Goldstein. I first met Hiley in 1998, in São Paulo, when he suggested I write a second biography of David Bohm. I balked at the suggestion but it stayed at the back of my mind. In 2015, when Angela Lahee, editor at Springer, suggested I write a biography of one of the quantum dissidents, Hiley's suggestion came back to the forefront and this time I did not hesitate. During all these years Hiley has read and commented on the papers I have written and spared time every time I have visited Birkbeck to talk with me; I have no words to express my gratitude for his kindness. While I did not record an interview, I also met Jeffrey Bub for several conversations about his experience as a student, first, and then a colleague of David Bohm's. I also had useful unrecorded conversations with Marco C. B. Fernandes and Jayme Tiomno. I am also indebted to other people whose remarks on Bohm or on my previous work on Bohm were for me intellectually challenging. They are Paul Forman and Michel Paty, and, in memoriam, Sam Schweber, Joan Bromberg, Alberto da Rocha Barros, and Amelia Hamburger.

Acknowledgements I have been working on Bohm's ideas since my Ph.D., obtained in 1995 at the University of São Paulo under the supervision of Michel Paty and Shozo Motoyama. My doctoral dissertation was dedicated to the analysis of Bohm's interpretation of quantum theory and its reception in the 1950s. Most of my time since then I have worked on the research which led to the book, *The Quantum Dissidents*, where Bohm was a major player. Thus, it is hard to list all the people and institutions I am grateful to for their support which resulted in this biography. While risking omissions, I would like to mention a few, in addition to those in the previous paragraphs. For the last sprint in the work on this biography, I obtained a leave of absence from my university, the Universidade Federal da Bahia, in Brazil. I am particularly grateful to João Carlos Salles, the president of the university, for understanding my need for leave of absence from my duties so as to finish this book. I spent this time at the American Institute of Physics, in its Center for History

---

[18]Wilkins (2003, 82). For the catalogue of Maurice Wilkins Papers, deposited at King's College, London, see: http://www.kingscollections.org/catalogues/kclca/collection/w/wilkins-maurice/. See, in particular, the following references: "K/PP178/12/26/1-23 Papers, 1965–1999, relating to the writings of quantum physicist and philosopher David Joseph Bohm (1917–1992)"

of Physics and Niels Bohr Library and Archives, in College Park, MD, in Washington, DC, area. I am thankful to its staff and particularly to Gregory Good, its director, and Stephanie Jankowski. During this time this study was financed in part by the Coordenação de Aperfeiçoamento de Pessoal de Nível Superior—Brasil (CAPES)—Finance Code 001 and the research was supported by the CNPQ [Grant 443335/2015-0]. I am grateful to Alex Wellerstein, for reading and commenting on Chaps. 2 and 3; Chris Talbot, for a careful reading of the full manuscript; Adam Becker, for sharing with me information about Bohm's FBI files, which he obtained through a FOIA request; Gustavo Rocha and Mirella Vieira, for discussions on Krishnamurti; my colleagues and students at the Laboratory Science as Culture [LACIC, Portuguese acronym], for discussion on biographies in the history of science; Cory Fischer, from the Krishnamurti Foundation; Amit Hagar, Amiram Ron, Gil Lainer, Yvette Gershon, Elaine Fletcher, Michael Liss and Nel Ben Ami, for obtaining documents and testimonies from Bohm's times at Technion and translation from the Hebrew; the Physics Institute at the Universidade de São Paulo and Ivã Gurgel, for the invitation to present a preliminary result of this book; Denise Sara Key, for her enduring support in the English revision; Angela Lahee for the suggestion and careful revision of the book proposal, as well as for her tolerance with my delays; Victoria Florio, for our conversation about science fiction magazines; Ricardo Zorzetto, for questions concerning the book; Italo Carvalho, for helping me with the graphs concerning scientometry; Thiago Hartz and Christian Joas, for their comments on a talk at the 25th International Congress of History of Science, held in Rio de Janeiro, 2017, which was an earlier version of Chap. 8; Monique Grimord, for discussions about the Pennsylvania mining towns; Chris Talbot, for several discussions on Bohm's work; Inés Cortazzo and Sonia Cabeda, for informal discussions about this project. My journey to Wilkes-Barre was easier thanks to the support of Peter Grimord, who also took some of the pictures over there; Agnes Soares, who explained the role of the Yearbook as source of information concerning American high schools, and Colleen Robatin and Patrick Peter, principals at the GAR Memorial Junior Senior High School. While I am thankful for all comments from these colleagues, undoubtedly however, faults in the final version of the manuscript are my entire responsibility.

I am also indebted to my sisters, Fatima, Inês and Silvana, and my son, Vitor, for their moral support; and Agnes Soares for her lovely and kind support and interest in this work.

References

Aharonov, Y., Bohm, D.: Significance of electromagnetic potentials in the quantum theory. Phys. Rev. **115**(3), 485–491 (1959)

Badino, M.: What have the historians of quantum physics ever done for us? Centaurus **58**(4), 327–346 (2016)

Bell, J.S.: On the Einstein Podolsky Rosen paradox. Physics **1**, 195–200 (1964)

Bell, J.S.: On the problem of hidden variables in quantum mechanics. Rev. Mod. Phys. **38**(3), 447–452 (1966)

Bell, J.S.: On the impossible pilot wave. Found. Phys. **12**(10), 989–999 (1982)

Bell, J.S.: Beables for quantum field theory. In: Bohm, D., Hiley, B.J., Peat, F.D. (eds.) Quantum Implications: Essays in Honour of David Bohm, pp. 227–234. Routledge & Kegan, London (1987)

Bohm, D.: Quantum Theory. Prentice-Hall, New York (1951)

Bohm, D.: A suggested interpretation of the quantum theory in terms of hidden variables—I & II. Physical Review. **85**(2), 166–179 (1952). (180–193)

Bohm, D.J., Hiley, B.J.: Intuitive understanding of nonlocality as implied by quantum-theory. Found. Phys. **5**(1), 93–109 (1975)

Bohm, D., Hiley, B.J.: The undivided universe: an ontological interpretation of quantum theory. Routledge, London (1993)

Bohm, D., Peat, F.D.: Science, Order & Creativity. Routledge, London (1987)
Bohm, D., Hiley, B.J., Peat, F.D.: Quantum Implications: Essays in Honour of David Bohm. Routledge & Kegan Paul, New York (1987)
Bohm, D., Biederman, C.J., Pylkkänen, P.: Bohm-Biederman Correspondence. Routledge, London (1999)
Carvalho, R.F.: Além das nuvens e dos relógios: A ideia de ciência de David Bohm e de Ilya Prigogine, Universidade Federal de Goiás [Brazil]. Ph.D. Dissertation (2015)
Caute, D.: Le Communisme et les intellectuels français, 1914–1966. Gallimard, Paris (1967)
Cross, A.: The crisis in physics: dialectical materialism and quantum theory. Soc. Stud. Sci. 21, 735–759 (1991)
Cushing, J.: Quantum Mechanics—Historical Contingency and the Copenhagen Hegemony. The University of Chicago Press, Chicago (1994)
Cushing, J.: Review of F. David Peat. Infinite potential: the life and times of David Bohm. Physics Today. 77–78 (1997)
Forman, P.: Weimar culture, causality, and quantum theory, 1918–1927: adaptation by German physicists and mathematicians to a hostile intellectual environment. Hist. Stud. Phys. Sci. 3, 1–115 (1971) Reprinted in Forman, P. et al. (eds.) Weimar Culture and Quantum Mechanics: Selected Papers by Paul Forman and Contemporary Perspectives on the Forman Thesis, Imperial College & World Scientific, London (2011)
Forstner, C.: The early history of David Bohm's quantum mechanics through the perspective of Ludwik Fleck's thought-collectives. Minerva 46(2), 215–229 (2008)
Freire Junior, O.: Science and exile: David Bohm, the cold war, and a new interpretation of quantum mechanics. Hist. Stud. Phys. Biol. Sci. 36(1), 1–34 (2005)
Freire Junior, O.: Quantum dissidents: research on the foundations of quantum theory, circa 1970. Stud. Hist. Philos. Mod. Phys. 40, 280–289 (2009)
Freire Junior, O.: Continuity and change: charting David Bohm's evolving ideas on quantum mechanics. In: Krause, D., Videira, A. (eds.) Brazilian Studies in Philosophy and History of Science, pp. 291–299. Springer, Heidelberg (2011)
Freire Junior, O.: The quantum dissidents: rebuilding the foundations of quantum mechanics (1950–1990). Springer, Berlin (2015a)
Freire Junior, O.: From the margins to the mainstream: foundations of quantum mechanics, 1950–1990. Ann. Phys. 5–6, A47–A51 (2015b)
Freire Junior, O., Silva, I.: Scientific exchanges between the United States and Brazil in the Twentieth Century: cultural diplomacy and transnational movements. In: Krige, J. (ed.) How Knowledge Moves—Writing the Transnational History of Science and Technology, pp. 281–307. Chicago University Press, Chicago (2019)
Freire Junior, O.: David Bohm e a controvérsia dos quanta. Centro de Lógica, Epistemologia e História da Ciência, Campinas [Brazil] (1999)
Gaddis, J.L.: The Cold War: A New History. Penguin, New York (2005)
Goldstein, S.: A theorist ignored: review of infinite potential. The life and times of David Bohm. In: David Peat, F.(ed.) Science, vol. 275, pp. 1893–1894 (1997)
Hobsbawm, E. (ed).: História do Marxismo [12 volumes]. Paz e Terra, Rio de Janeiro (1991)
Hobsbawm, E.: Age of Extremes—The Short Twentieth Century 1914–1991. Abacus, London (1995)
Hobsbawm, E.: Interesting Times: A Twentieth-Century Life. Pantheon Books, New York (2002)
Hobsbawm, E.: How to Change the World: Reflections on Marx and Marxism. Yale University Press, New Haven, Conn (2011)
Jammer, M.: The Philosophy of Quantum Mechanics—The Interpretations of Quantum Mechanics in Historical Perspective. Wiley, New York (1974)
Judt, T.: Postwar—A History of Europe Since 1945. Penguin, New York (2005)
Kojevnikov, A.: Review of F. David Peat. infinite potential: the life and times of David Bohm. ISIS. 89(4), 752–753 (1998)
Kojevnikov, A.: David Bohm and collective movement. Hist. Stud. Phys. Biol. Sci. 33, 161–192 (2002)

Kožnjak, B.: The missing history of Bohm's hidden variables theory: The ninth symposium of the
    Colston research society, Bristol, 1957. Stud. Hist. Philos. Mod. Phys. **62**, 85–97 (2018)
Kragh, H.: An Introduction to the Historiography of Science. Cambridge University Press, Cam-
    bridge (1987)
Krige, J. (ed.): How Knowledge Moves—Writing the Transnational History of Science and Tech-
    nology. Chicago University Press, Chicago (2019)
Le Goff, J.: Saint Louis. University of Notre Dame Press, Notre Dame, IN (2009). (transl. by G. E.
    Gollrad)
Moody, D.E.: An Uncommon Collaboration—David Bohm and J. Krishnamurti. Alpha Centauri
    Press, Ojai, CA (2017)
Mullet, S.K.: Bohm, David Joseph. New Dictionary of Scientific Biography. N. Koertge. New York,
    Thomson—Gale. I, pp. 321–326 (2008b)
Mullet, S.K.: Little man: four junior physicists and the red scare experience. Harvard University.
    Ph.D. Dissertation (2008a)
Olwell, R.: Physical isolation and marginalization in physics—David Bohm's cold war exile. ISIS
    90, 738–756 (1999)
Ory, P., Sirinelli, J.-F.: Les Intellectuels en France de l'affaire Dreyfus à nos jours. Perrin, Paris
    (2004)
Peat, F.D.: Infinite potential: the life and times of David Bohm. Addison Wesley, Reading, Mass
    (1997)
Petrônio, R.: Apresentação da tradução, in Bohm, D. Causalidade e Acaso na Física Moderna,
    [Portuguese translation, with presentation and notes by R. Petrônio]. 9–42, Contraponto, Rio de
    Janeiro (2013)
Philippidis, C., et al.: Quantum interference and the quantum potential. Nuovo Cimento della Soci-
    eta Italiana di Fisica B-General Physics Relativity Astronomy and Mathematical Physics and
    Methods **52**(1), 15–28 (1979)
Pinch, T.: What does a proof do if it does not prove? A study of the social conditions and metaphysical
    divisions leading to David Bohm and John von Neumann failing to communicate in quantum
    physics. In: Mendelsohn, E., Weingart, P., Whitley, R. (eds.) The Social Production of Scientific
    Knowledge, pp. 171–216. Reidel, Dordrecht (1977)
Dosse, F.: Biographie, prosopographie. In: Delacroix, C., Dosse, F., Garcia, P., Offenstadt, N. (eds.)
    Historiographies: concepts et débats, vol 1, pp. 79–85. Gallimard, Paris, (2010)
Pylkkänen, P.: Mind, Matter and the Implicate Order. Springer, Berlin (2007)
Schrecker, E.: No Ivory Tower: McCarthyism and the Universities. Oxford University Press, New
    York (1986)
Schrecker, E.: The Age of McCarthyism—A Brief History with Documents. Bedford/St. Martins's,
    Boston (2002)
Seth, S.: Crafting the Quantum—Arnold Sommerfeld and the Practice of Theory, 1890–1926. The
    MIT Press, Cambridge, Ma (2010)
Seth, S.: The History of Physics after the Cultural Turn. Hist. Stud. Nat. Sci. **41**(1), 112–122 (2011)
Söderqvist, T.: Existential projects and existential choice in science: Science biography as an edi-
    fying genre. In: Shortland, M., Yeo, R. (eds.) Telling Lives in Science—Essays on scientific
    biography, pp. 45–84. Cambridge University Press, Cambridge (1996)
Staley, R.: Trajectories in the history and historiography of physics in the twentieth century. Hist.
    Sci. **51**, 151–177 (2013)
Talbot, C.: David Bohm: Causality and Chance, Letters to Three Women. Springer, Berlin (2017)
Wang, J.: American Science in an Age of Anxiety: Scientists, Anticommunism, And the Cold War.
    University of North Carolina Press, Chapel Hill, NC (1999)
Westad, O.A.: The Cold War—A World History. Basic Books, New York (2017)
Westfall, R.S.: Never at Rest: A Biography of Isaac Newton. Cambridge University Press, New
    York (1980)
Wilkins, M.H.F.: The Third Man of the Double Helix: The Autobiography of Maurice Wilkins.
    Oxford University Press, Oxford (2003)

Chapter 2
From Wilkes-Barre to a Physics Ph.D. at Berkeley (1917–1945)

In 1941, David Joseph Bohm moved from Caltech, where he had arrived in 1939, to the University of California at Berkeley to advance his doctoral degree under the supervision of Julius Robert Oppenheimer. Earlier, at high school in Wilkes-Barre, PA, and at the Pennsylvania State College, he had already exhibited an outstanding talent for mathematics and science. When he arrived at Berkeley, he had already consolidated two intellectual trends which would characterize his whole life. He knew what kind of physics he enjoyed working on; while he was attracted to theoretical physics and had exhibited skills in mathematics, he had no patience for the solving-problem style of physics he had found at Caltech. Instead he looked for speculative and conceptual science. Bohm, however, was not only concerned with science, for him science was part of a larger picture involving society as a whole. During the Great Depression he had shifted from a strong commitment to individualism along the lines of the American Dream to a more socially inclined, even sympathetic, social view. At Berkeley these preferences acquired concrete forms. Bohm was able to work on different subjects, all of interest to the war effort, and obtain his doctoral degree. Still, his social views led him to join the Communist Party. In the first section of this Chap. 1 analyze his early life, including the formative experiences at family, high school, and college as well as the context in which his social awareness began to flourish. Then I discuss the formation of Bohm's scientific style, which matured in the distinct experiences at Caltech and at Berkeley. In the third and final Sect. 2.1 analyze his years at Berkeley, both for his political engagement and his scientific training and works.

2.1 Before Arriving at Berkeley

David Bohm first became fascinated by science in his early teens reading the science fiction magazine *Amazing Stories*, which had recently become popular in the US. Bohm kept a vivid recollection of this first reading: "I think the first interest [in

© Springer Nature Switzerland AG 2019
O. Freire Junior, *David Bohm*, Springer Biographies,
https://doi.org/10.1007/978-3-030-22715-9_2

science] that I can remember was when I was about eight-years-old and in my father's store. There was a boy, he used to have boys come into help him take care of it. He brought a magazine called *Amazing Stories*, a science fiction journal. [...] It had a story called 'The Columbus of Space' or something or 'Voyage to Venus.' I don't remember what the story was, but this tremendously aroused my interest, you see". The article Bohm read was "A Columbus of space," a three-part story of a rocket journey into outer space, namely to Venus, which had been written by Garrett Putnam Serviss, an astronomer, popularizer of astronomy, and science-fiction writer. It began to be published in the very first volum[1]e, issue 5, of *Amazing Stories*, in 1926.

Stories of space travel were not new, we just need to recall Julio Verne's 1865, *From the Earth to the Moon*. What was new was that they became the bread and butter of American science-fiction popular magazines for the young audience. These travels also stirred up debates as there was no known power to launch such endeavors. The editor of *Amazing Stories*, Hugo Gernsback, fought back, "For centuries the human mind has groped with the problem of soaring into space and exploring other worlds. [...] only in the last thirty years mechanical flight has been accomplished," thus considering conservative those thinkers who did not entertain such possibilities.[2] Bohm was thus among the first to be seduced by this new cultural trend in the US.

This escape from daily life was also motivated by the grim family environment in which Bohm grew up. The eldest in his family, he was born on December 20, 1917 in Wilkes-Barre, Pennsylvania, son of Samuel and Frieda Bohm, both Jewish immigrants, of Hungarian and Lithuanian origin, respectively. Samuel Bohm ran a second-hand furniture store at 410 Hazle Street, which he had taken over from his father-in-law, who had hosted Samuel when he arrived in the U.S. as a teenage immigrant. Frieda Bohm suffered from mental instability since arriving in the US, when she had difficulty coping with the new language and environment. Thus it is plausible to think theirs was a marriage of convenience. Due to her mental outbursts, Frieda could not cope with running the home and Samuel was constantly irritated and aggressive towards her; according to F. David Peat, Bohm's first biographer, "The home into which David was born was chaotic, oppressive and at times, violent". On the one hand, Bohm sustained affection for his mother but did not feel sure that she could take care of him. On the other hand, he did not identify himself with his father who did not appreciate his inclination for science. Bohm grew up, according his later recollections, without a wide circle of friends, almost an introspective person, partly due to this family ambiance but also from an aversion to sport. However, he did enjoy walking, hiking, and talking with the few friends he had about his musings on science and social issues.[3]

---

[1] Interview of David Bohm by Maurice Wilkins on 1986 June 6, Niels Bohr Library and Archives, American Institute of Physics, College Park, MD USA, www.aip.org/history-programs/niels-bohr-library/oral-histories/32977-1. G. P. Serviss, A Columbus of Space, Amazing Stories, 1(5), 388–409; 474–475; 1(6), 490–509; and 1(7), 596–615 (1926). I am indebted to Victoria Florio de Andrade for unearthing the article and discussing its context. On the role of popular science and science-fiction magazines in the US in the 1920s, see (Andrade 2017).

[2] See Gernsback (1927).

[3] Peat (1997, 13).

If the family environment was grim, Bohm had more fun at school. He attended the G. A. R. Memorial Junior/Senior High School, located at 250 South Grant Street. There he blossomed both as a promising science student and as a teenager with good social skills. Bohm's persona among his fellow high school colleagues was recorded in the 1935 Yearbook, on the occasion of his graduation. His calling and talent for science and mathematics is evidenced in his nickname, "Einstein," as well as in the manner colleagues and teachers portrayed his personality: "Deep sighted in intelligences, Ideas, atoms, influences". From this publication Bohm's other social skills can be inferred. He was part of the "Commencement Announcement Committee," which testifies his popularity among colleagues, and attended the German Club, the Math Club, two Dance committees, and Swimming Club. From later recollections, we also know how impressed Bohm was by learning geometry guided by his Mathematics teacher, Meyer Tope.[4]

The memory of Bohm is alive at the current G. A. R. High School. In 2004, the school created the "David Bohm Science Award," for excellence in science and mathematics. In the plate with the list of the students who have been awarded, it is written: "David Bohm was a 1935 graduate of G. A. R. Memorial High School. He was one of the most brilliant minds of the Twentieth Century. A world renowned physicist explorer of consciousness and colleague of Albert Einstein". If we consider the formative experiences Bohm had at home and at school, we may note, incidentally, that David Peat's focus on Bohm's family troubles led him to underestimate the positive role played by school in Bohm's education and personality (Figs. 2.1, 2.2, 2.3, 2.4, 2.5 and 2.6).

2.1.1 The Inception of Concerns with the Big Social Picture

Bohm's concerns with politics and societal issues were awakened still as a teenager stimulated by the social environment in Wilkes-Barre. This Pennsylvania coal mining town is part of the larger Appalachian anthracite coal region where the extensive and intensive coal mining began in the mid 19th century. Working conditions were oppressive, with the widespread use of child labor and unfair subcontracting systems, and labor unions and battles were part of the daily life of the region. However, Wilkes-Barre was a flourishing town with the coal and garment industries and a population of over eighty thousand people. In the 1920s and 1930s, this economic growth stalled, first due to the changes in the use of fuel with the decline of anthracite for home

---

[4]The GARCHIVE 1935, published by the Senior Class of G.A.R. Memorial High School, Wilkes-Barre, Pennsylvania. Recollections about the mathematics teacher, Meyer Tope, are in Interview of David Bohm by Maurice Wilkins on 1986 June 12, Niels Bohr Library and Archives, American Institute of Physics, College Park, MD USA, www.aip.org/history-programs/niels-bohr-library/oral-histories/32977-2. In the transcription of this interview his name is spelled as Mario Tope. The transcription was not revised by Bohm himself and the school records indicate the right name is Meyer Tope. I am thankful to Colleen Robatin for facilitating my consultation of the school records and to Agnes Soares for discussing the meaning of some of these records.

Fig. 2.1 A Columbus of
Space, by G. P. Serviss, the
science-fiction story which
fascinated Bohm in his
childhood, published in
Amazing Stories, 1926

heating purposes, and then from the vicissitudes of the Great Depression. The crisis
brought huge social unrest and unemployment to the region. Indeed, the decline
on the US mining towns in that region would become a lasting drama, with ups
and downs, which included few periods of recovery succeeded by new declines.
Nowadays the signs of decline are omnipresent in Wilkes-Barre with a population
around forty thousand people. The most recent chapter in this drama took the shape
of a clash in the US 2016 presidential elections between the two major candidates
presenting different proposals for the future of the region.[5]

Back to the 1930s, we have a vivid testimony from somebody who lived most of the
1930s in Wilkes-Barre trying to organize the unemployed in the fight for their rights.
"Behind the very real beauty of the region lay a bitter poverty. The depression had
started there here in the twenties, and by the time it reached other areas, the eastern

---

[5]On the history of labor contracts and battles in the anthracite region, see Wolensky and Hastie
Sr. (2013). On the clash in the presidential elections concerning the drama of the mining towns,
see the NPR podcasts "In The Coal Counties Of Central Appalachia, Will Trump's Promises Come
True?" 09 May 2018, available at https://www.npr.org/2018/05/09/607273500/in-the-coal-counties-
of-central-appalachia-will-trump-s-promises-come-true, accessed on 23 Jan 2019. I am thankful to
Monique Grimord for discussion on this subject.

Fig. 2.2 410 Hazle Street, Wilkes-Barre, PA, where Bohm's father ran his second-hand furniture store and lived on the first floor. As an evidence for the economic decline of the town, the current store went out of business. *Credits* Picture by Peter Grimord

Pennsylvania mining towns were already devasted". The impact of the depression on the working classes at the towns may be illustrated by the following figure: "the number of miners in the industry fell from 140,000 in 1928 to less than 80,000 in 1930". At that time there was no unemployment benefits. This dramatic social crisis created social unrest and the workers began to organize themselves to protest. They were helped by field organizers from unions and left groups such as the Communists. These grassroot organizations were not accepted by the owners of the coal companies. Still according to this testimony, in Wilkes-Barre, "the political repression was as bad as anywhere in Illinois, and the economic situation was much worse," and "there was virtually a rule of terror by the coal companies, who controlled the police and local officials in most towns". However, the leaders of the social movement also found some support in the Wilkes-Barre's local elites, which included some teachers, ministers, lawyers, and even the mayor. The picture in Fig. 2.10 taken from a 1933 manifestation in Wilkes-Barre, is a clear evidence of the strength of this movement. Still according to this testimony, the movement of the unemployed in this region contributed to change the way the unemployment was treated in the US. According to him, "in the summer of 1935, Congress passed the Social Security Act. This bill incorporated our main goal of unemployment compensation as well as a pension system".[6]

---

[6]Nelson et al. (1981, 94–161).

GAR Memorial High School is located in the Heights section of the city on Grant and Lehigh
Streets. Classes began here in 1925, and contrary to popular belief, the school's initials stand
for the Grand Army of the Republic. Students attending this school reside in the eastern part
of the city.

Fig. 2.3 G.A.R. Memorial Junior Senior Highschool, Wilkes-Barre, PA. Façade in the 1930s, when
Bohm studied there. *Credits* Elena Castrignano, Wilkes-Barre, Postcard History Series, Arcadia
Publishing, 2008

Fig. 2.4 G.A.R. Memorial Junior Senior Highschool, Wilkes-Barre, PA. Current façade. *Credits*
Picture by Peter Grimord

Fig. 2.5 Bohm portrayed in the Yearbook 1935 from the G.A.R. Memorial Junior Senior High-school, Wilkes-Barre, PA. Interestingly Bohm was nicknamed by his colleagues, Einstein, and portrayed as "deep sighted in intelligences, ideas, atoms, influences". *Credits* The GARCHIVE (1935), courtesy of Colleen Robatin

The authorship of this testimony has some overlapping with Bohm's life history. These recollections came from the autobiography of Steve Nelson, who was a member of the Communist Party and spent most of the thirties organizing coal workers in the region of Wilkes-Barre. He only left the town to voluntarily fight in the Spanish Civil War, coming back to the town for a while after this.[7] Bohm's and Nelson's life would become entangled during the war, at Berkeley, in an episode which would be later used against Bohm during McCarthyism. We will come back to this later in the book. As far as we were able to track down, neither Nelson nor Bohm recalled having met each other while Bohm was a teenager living in Wilkes-Barre. In fact, a later research conducted in 1943, when Bohm was already at Berkeley, led the FBI to conclude that "no derogatory information relative to either the subject [Bohm] or his family was developed at Wilkes-Barre concerning Communistic or other radical tendencies".[8]

"My main interest was really physics you see. Although I had a vivid interest in politics and the general state of civilization;" these were Bohm's recollections, more than forty years later, from his high school and college times. While we should take any later recollections with a pinch of salt, as they are also shaped by current experiences, it is appealing to follow Bohm's recollections of his political and cultural views as they evolved throughout his life. As a teenager Bohm strongly believed in the American dream as consequence of individual achievements. As the Great Depression hit Wilkes-Barre, individualism ceded room for more socially and collectively inspired views. Towards the end of his high school years the news from

---

[7]Nelson et al. (1981).

[8]David Bohm's FBI File, (1358423-0-100-HQ-207045, p. 7). I am grateful to Adam Becker for sharing with me these documents which were obtained through a FOIA [Freedom of Information Act] request.

Fig. 2.6 Bohm's Prize for outstanding students in science and mathematics was created by the G.A.R. Memorial Junior Senior Highschool, Wilkes-Barre, PA. It has been awarded since 2004. *Credits* Picture by O. Freire

Fig. 2.7 Child labor was still common in the coal mines at the time Bohm was born. Title of this picture: Group of Breaker boys. Smallest is Sam Belloma, Pinc Street. Location: Pittston, Pennsylvania. In 2018 these pictures were part of the exhibition Anthracite Photographers: Photographers of Anthracite, Anthracite Heritage Museum, Scranton, PA, curated by Black et al. (2018). *Credits* Picture by Lewis Hine. Library of Congress, Prints and Photographs Division, National Child Labor Committee Collection, LC-DIG-nclc-01137

Europe added anxiety to the American ambiance. Bohm was against Mussolini while his father thought Mussolini would assure order in Italy and make Italian trains run on time. His concerns with the political context in Europe grew as the Nazi took power in Germany, "I sort of felt Mussolini was a contemptible person but Hitler was really dangerous," these are Bohm's own reminiscences.[9]

Bohm's social concerns were nurtured while in Wilkes-Barre reading socialist newspapers, such as *The Nation* and *New Republic*, some of which were available at the local public library. As a landmark for the ulterior evolution of his political views, at high school Bohm wrote a short sketch criticizing both Hitler and Stalin and at college he did not join radical student groups because he found them too pro USSR. Criticisms towards the Soviets would be watered down as the war evolved in Europe. Indeed, at the end of his first year at Caltech, back to spend the summer at home, Bohm was gloomy about the future. He felt powerless as the Nazis had invaded France and Holland and did not see any will in the US to fight them. The ideals of freedom and individualism in his home country were just lip service, as his fellows were not committed to fight for them, these were his recollections of his feelings at the time. His morale would dramatically change at Berkeley (Figs. 2.7, 2.8, 2.9, 2.10 and 2.11).[10]

---

[9]Interview of David Bohm by Maurice Wilkins on 1986 June 12, AIP, College Park, MD USA, www.aip.org/history-programs/niels-bohr-library/oral-histories/32977-2.

[10]Interview of David Bohm by Maurice Wilkins on 1986 June 12 and July 7, AIP, College Park, MD USA. The full interview is available at https://www.aip.org/history-programs/niels-bohr-library/oral-histories.

Fig. 2.8 Child labor was still common in the coal mines at the time Bohm was born. Title of this picture: Breaker boys in #9 Breaker, Hughestown Borough, Pa. Coal Co. The smallest boy is Angelo Ross, Location: Pittston, Pennsylvania. In 2018 these pictures were part of an exhibition of Anthracite Photographers: Photographers of Anthracite, Anthracite Heritage Museum, Scranton, PA, curated by Black et al. (2018). *Credits* Picture by Lewis Hine. Library of Congress, Prints and Photographs Division, National Child Labor Committee Collection, LC-DIG-nclc-01139

2.2   The Shaping of a Scientific Style

During high school in Wilkes-Barre, Bohm nurtured a combination of highly imaginative ideas either in science-fiction or in science itself with the acquisition of mathematical skills. From this time Bohm had recollections of being impressed by tornados, trying to see how objects and shapes could emerge from sole movements. He had a vivid recollection of a hurricane and its consequences hitting his father's store. And yet, as we have seen, his first contact with geometry left an indelible mark.[11]

High school over, his father supported him to go to Pennsylvania State College, near Wilkes-Barre (120 miles away), despite mistrusting science as a potential professional career. Penn State, now a university with 100,000 students and 24 campus, had been set up in the mid 19th century, and was then a college with a stronghold position in agricultural sciences but with a small number of students doing undergraduate degrees in physics. Indeed, as Bohm recalled later, it had about five physics students when he was there.

---

[11] Interview of David Bohm by Maurice Wilkins on 1986 June 12, AIP, College Park, MD USA, www.aip.org/history-programs/niels-bohr-library/oral-histories/32977-2.

Fig. 2.9 Campaign against the child labor in the coal region at the end of the 1910s. Title: Exhibit Panel, c. 1913 or 1914. In 2018 these pictures were part of the exhibition of Anthracite Photographers: Photographers of Anthracite, Anthracite Heritage Museum, Scranton, PA, curated by Black et al. (2018). *Credits* Picture by Lewis Hine. Library of Congress, Prints and Photographs Division, National Child Labor Committee Collection, LC-DIG-nclc-04924

Penn was not an enticing intellectual environment for a would-be physicist. Quantum mechanics teaching, for instance, was limited to Bohr's atom. In spite of these constraints, the four years Bohm spent at Penn were very positive for his training in physics. He attended a good mathematics course, dealing with topics such as analytic functions, series, and transcendental functions and solving problems from *A course of modern analysis*, a standard university textbook written by E. T. Whittaker and G. N. Watson, whose first edition was in 1902. Later on, at Caltech, he realized how appropriate the mathematics and physics skills he had acquired at college were for physics research. In addition, he had time to constantly walk around the town, a college town, and talk to colleagues about science and social issues. This lifestyle, walking and talking, would become his favorite way of thinking about scientific issues.[12]

---

[12]Whittaker and Watson (1943).

A meeting of the unemployed in Wilkes-Barre, 1933.

Fig. 2.10 Meeting of unemployed miners, in Wilkes-Barre, 1933. *Credits* Nelson et al. (1981)

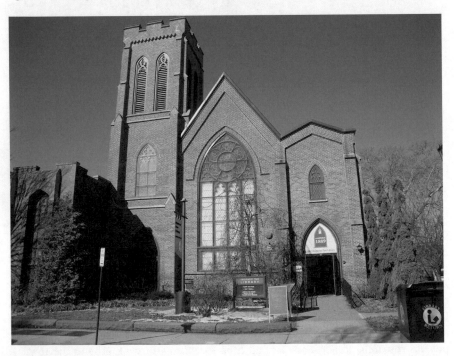

Fig. 2.11 Osterhout Free Library, Wilkes-Barre, PA, where Bohm used to read the magazines The Nation and New Republic and developed his social awareness. *Credits* Picture by Agnes Soares

In the last year at Penn Bohm looked for a place to do a doctoral degree in physics. He applied to different universities but failed to get positive answers when two different chains of events occurred. He had a positive answer from Caltech offering him tuition and what he had longed for since a teenager, to move West. Thanks to his good results he won a $600 fellowship from Penn to pursue doctoral studies in a place of his choice. Thus, at the end of the summer of 1939, September the 1st, he took a bus to Pasadena, California. On the same day, Hitler invaded Poland, starting the war in Europe. Bohm embarked on his journey apprehensive but the subject did not spur much conversation on the trip.

At Caltech Bohm's inclinations towards a scientific style more based on concepts and fundamentals while not discarding the mathematical machinery came of age. He had no difficulty solving the problems on the electromagnetism course, which was given by William Ralph Smythe, based on his textbook *Static and dynamic electricity*, but Bohm soon became bored with this course. Smythe's course was based on his 12-years of experience teaching and it followed the tradition of a detailed presentation of mathematical methods to deal with electrostatic and electrodynamic problems only introducing the full theory of electromagnetism, Maxwell's equations, after 80% of the book had been covered. Smythe's approach, however, was not a naïve pedagogical choice. Instead, it was intended to fill a gap he had identified in the students' ability to solve problems. As he stated in the opening of his book, "It has been found that, in most cases, the average graduate student, even though he seems to be thoroughly familiar with advanced electrical theory, is unable to solve the electrical problems he encounters in research if they fall outside the routine types and so must be worked out from first principles". Thus, "more than the usual number of problems have been worked in the text". Smythe had no doubt about the kind of scientist he wanted to train with such a book and course: "This book is written for the experimental research physicist and engineer rather than for the theoretical man".[13]

Smythe's goals collided with Bohm's expectations. The teacher perceived Bohm's discomfort but only promised a course exploiting theories for the following year. According to his recollections, Caltech and Smythe's course were not what he had expected. Later he would say "It wasn't what I had hoped it would be. It was just full of problems and we used to solve them". Another hint about the kind of scientific style Bohm was nurturing was his attraction to Eddington's 1936 book, *Relativity theory of protons and electrons*, which was kind a precursor to the current quest for a theory of everything. In fact, since the early 1920s Eddington had been looking for a grand unification between the general theory of relativity and electromagnetic theory. After the inception of quantum mechanics, and particularly after Dirac's equation for the relativistic electron, Eddington attempted to rewrite this equation in terms of tensor calculus in order to reconcile quantum mechanics and gravitation. According to historian Helge Kragh, Eddington's book "was as remarkably ambitious as it was

---

[13]Smythe (1939, v).

remarkably unsuccessful".[14] Bohm tasted both its ambition and lack of success. He was not able to understand the technicalities of the book due to its mathematical level but tried to write a paper presenting the main ideas of the book in a simplified manner. Bohm took the subject to discuss with Richard Tolman, a renowned Caltech professor, who rebuked it causing Bohm some frustration.

In the first year of graduate studies at Caltech students had neither research problems nor supervisors, they simply followed courses and solved problems. This was not to Bohm's liking. In addition, he found the intellectual ambiance among the students too competitive. Furthermore, he did not like the dry weather and the almost desert like vegetation around Pasadena. He could only relax hiking round Mount Wilson nearby. In his second year Bohm got an assistantship which covered his living expenses but he was not motivated to pursue the doctoral studies to their conclusion.

At that point, casual events helped Bohm. Julius Robert Oppenheimer commuted from Berkeley to Pasadena once a week to lecture. A mutual acquaintance, Milton Plesser, introduced Bohm to Oppenheimer and within a few weeks Bohm had moved to Berkeley and got an assistantship there. Arriving at Berkeley his first positive impression was the climate, which he found far better than in Pasadena.

What Bohm did not realize during his times at Caltech and at Berkeley under Oppenheimer was the existence of varying styles in physics, differences which reflect deeper entrenchment in the practice of this science throughout history. At this point one may wonder how the concept of style, so common in art, literature and humanities at large, may be related to science. Indeed, historians and philosophers of science have suggested that when one consider as sciences, physics in the case, not only their consolidated results but also the work implied in their production, the physicists need to make choices in their reasonings. These choices may be ruled by certain personal choices, which may also be shared by a group of scientists, thus creating a school or a tradition, and the understanding of these choices may be enlightened by the concept of style.[15] Bohm met at Caltech and at Berkeley two different styles. Not only are both styles grounded in the tradition in physics but both have been able to produce achievements in the evolution of science.

We may illustrate this with a few examples from the physics community which produced quantum mechanics in the first quarter of the 20th century. Arnold Sommerfeld and the school he organized in Munich was a first rank achievement of theoretical physics in Germany. If evaluated by his scientific achievements, Sommerfeld introduced the theory of relativity to Bohr's atomic model and by doing so he introduced new quantum numbers and the fine-structure constant in atomic spectra in addition to pioneer studies of X-rays with wave theory. Among the doctoral students he was able to train were Peter Debye, Max von Laue, Wolfgang Pauli, Werner Heisenberg, Linus Pauling, and Hans Bethe, all Nobel Prize winners.

---

[14]Interview of David Bohm by Maurice Wilkins on 1986 July 7, AIP, College Park, MD USA, www.aip.org/history-programs/niels-bohr-library/oral-histories/32977-3. Eddington (1936). Kragh (1999, 219). On Eddington's attempt, see also Kragh (2011, 92–101).

[15]On style in science, see Paty (1990), Chap. 4, and Paty (1993), Chap. 1.

All these achievements were consequences from Sommerfeld's own style of doing physics. The historian Suman Seth has studied this style thoroughly and called it the "physics of problems," a label which was not disliked by Sommerfeld himself. Later on, while writing one of his many influential textbooks, Sommerfeld acknowledged this by stating, "In print, as in my classes, I will not detain myself with the mathematical foundations, but proceed as rapidly as possible to the physical problems themselves". This style, according to Seth, is better understood when compared to the other major style, which was very influential in German theoretical physics in early 20th century. Seth tells us that "Sommerfeld, and others, saw his vision of physics—concerned with 'the physical problems themselves'—as qualitatively different from an alternative kind of physics, including Max Planck's or Albert Einstein's, which was concerned with subsuming all physical phenomena under a few abstracted, generalized axioms". Still according to Seth, we may speak of these two contrasting styles: "Thus, in contrast with the 'physics of principles' espoused by Planck, Einstein, and (to an extent) Niels Bohr, Sommerfeld's was a 'physics of problems'".[16]

If Sommerfeld illustrates the kind of solving-problems physics Bohm met at Caltech, Niels Bohr and Albert Einstein illustrate a quite different style, where concepts, pictures, and principles play a major role. Both styles, solving-problems and concepts and principles, flourished in American physics between the wars but the former gained more momentum. According to the historian of physics Sam Schweber, in his influential study "The empiricist temper regnant—theoretical physics in the United-States 1920–1950," the manner in which physics departments were organized in the US, bringing under the same roof theoretical and experimental physicists, contributed to shape the "empirical, pragmatic, instrumentalist" style of theoretical physics in the US.[17] The conflict between these two styles would follow Bohm throughout his professional life with an apex in the 1950s when he would suggest a new interpretation for quantum mechanics.

2.3 The Berkeley Years

In June 1941 Bohm settled in at Berkeley where he would have a life-changing experience both in terms of social engagement and sensibilities and his scientific training. At Berkeley, everything was friendlier than at Caltech. He had a charismatic Ph.D. supervisor, Oppenheimer, who was also a focal point for young left wing people; the weather was better and he had a problem—scattering of protons from deuterons—to work on. Oppenheimer played a paternal role for Bohm who developed both an affection and admiration for him. As Bohm would recall later—"he was going to fulfill the role of the father, which was not fulfilled". The role played by Oppenheimer in Bohm's life inspired Bohm's early biographer, F. David Peat, to

---

[16]Seth (2010, 2), including the citation of Sommerfeld, from his *Mechanics* (1943).
[17]Schweber (1986, 58).

appeal to the Freudian psychoanalytical framework to suggest that the trauma due to his dysfunctional family would follow Bohm through his entire life. There is no doubt that Oppenheimer had good social skills and intellectual strength in addition to impressive scientific capacities. This was the reason why Leslie Groves, the military officer in charge of the Manhattan project, would choose him to lead the full team of scientists assembled in Los Alamos. The success of this top secret US atomic project was also a result of Oppenheimer's leadership in spite of his later revealed character failure, which also hindered Bohm.[18] Let us now focus on Bohm's concerns about "politics and the general state of civilization," as he saw in hindsight his interests other than science.

2.3.1 Political Engagement

Bohm's engagement with the Communist Party happened at the confluence of two different contexts, which are related to the spread of World War II, on the one hand, and the Berkeley social and political atmosphere, on the other. To these contexts we should add his early social interests, which date, as we have seen, from his high school years. On December 7, 1941, the Japanese attack on the US military base in Pearl Harbor led the US to declare war on Japan entering thus the Second World War. The US, the UK and the USSR were now on the same side on the battlefield. This alliance also included a number of different circles of activists in the world, creating a totally new military situation in world affairs, which would lead the historian Eric Hobsbawm to create the oxymoron "international ideological civil war" against fascism to describe it.[19] Bohm, who was following the evolving war in Europe, concerned about the weakness of the Western reaction to Nazi expansion, greeted the US decision with enthusiasm.

In addition to that wider background, the other context which also contributed to Bohm joining the Communist Party was because he felt at home with Oppenheimer, his students and his acquaintances. The subjects discussed ranged from the international situation, technical physics problems to philosophical interpretations of quantum mechanics, readings of Marxist works, and engagement in unionizing technical and scientific workers at Berkeley. Bohm's closer colleagues at Berkeley in his

---

[18]Interview of David Bohm by Maurice Wilkins on 1986 July 7 and June 12, AIP, College Park, MD, USA. There is a literary industry on Oppenheimer and his role in the Manhattan project and later in the US government agencies dedicated to nuclear issues. It includes the canonical description of the project, written by Richard Rhodes (1986), and biographical works such as Bird and Sherwin (2005). For Oppenheimer's moral failures, later revealed, particularly concerning Bernard Peters and its impact on the physics community, see Schweber (2000, 115–130).

[19]Hobsbawm (1995, 144). For a summary analysis of the short-lived situation of convergence between the US and the USSR, see Hobsbawm (1995, 142–177) and his description of the context "Against the Common Enemy". However, he excluded regions under colonial rule, such as Africa, parts of Asia and the Far East, from this "civil war". The subject is further developed in Chapter 11 of (Hobsbawm 2011).

first years included the physicists Bernard Peters, Joseph Weinberg, Giovanni Rossi Lomanitz, and Frank Oppenheimer and all of them would suffer as a result of the anti-communist mentality which would take hold in American political circles later. Bohm's acquaintances and friendships also included physicists Philip Morrison and Melba Phillips, in addition to the psychologist Betty Friedan, who was Bohm's first girlfriend. While Bohm had the impression that Oppenheimer, his supervisor, was not very interested in discussing the war and the political situation with him, particularly after the US entered the war, and recalls going only once to a party at Oppenheimer's home, other impressions may be obtained from the description of group activities. Later in the 1950s, Bohm described these experiences to the French physicist and member of the French Communist Party Jean-Pierre Vigier, who would become a close associate of Bohm in the works on the interpretation of quantum mechanics. As a consequence of Bohm's descriptions, "Vigier was given the impression of an active Communist cell".[20]

As a matter of fact, in November 1942, according to David Bohm's later recollections, he joined the Communist Party. However, it is hard to say how long he kept this connection and how far his engagement went. His own later testimonies are not entirely consistent. As noted by his first biographer, F. David Peat, according to Bohm's interview with Martin Sherwin, "the meetings were so boring, [that] he dropped out after only a few weeks". Peat continues, "yet in a statement he later made to the American consul in London, it appears that he remained a party member for nine months". This is not surprising from somebody who was a target of the anti-communist campaigns, later known as McCarthyism, which dominated the American political scene from the late 1940s on until after the onset of the Cold War. These circumstances would lead him to lose his job in the US, live in exile, and to lose his American citizenship for almost 30 years.[21]

However, if details of Bohm's engagement with the Communist Party are not clear, we can be sure of two things. First, he began to suffer the effects of his political engagement very soon after. His attempts to organize at Berkeley Radiation Laboratory a local chapter of the Federation of Architects, Engineers, Chemists and Technicians, the FAECT, a CIO-sponsored union, were monitored by the US intelligence in the context of the beginning of the atomic project, where Berkeley physicists would play a huge role. Berkeley Laboratory was the place where Ernest O. Lawrence had invented the cyclotron, one of the earliest subatomic particle accelerators, leading him to win the 1939 Physics Nobel prize. When the top secret American atomic project was launched this accelerator was used to design a method for enriching uranium, which was a difficult and the largest part of the project. This method, embodied in a new device, calutron, was eventually brought to an industrial scale to produce the fissile uranium used in the bomb. According to historian Richard Rhodes, "making the new instrument work, through the spring and summer of 1942, solved the most difficult design problems. It acquired a name along the way: calutron, another tron

---

[20]Interview of David Bohm by Maurice Wilkins on 1986 July 7, AIP, College Park, MD, USA. Peat (1997, 57).

[21]Bird and Sherwin (2005, 172); Peat (1997, 58).

from the University of California". In addition, Oppenheimer was invited by Leslie Groves to be the scientific director of the project. To give a flavor of the complexities of the time, Oppenheimer himself, who would become the leader of the atomic project, had helped the organization of this union chapter.[22]

As we now know security officials saw union activities as part of Communist activities and monitored Bohm throughout. Evidence, only decades later declassified, shows us how high ranks officials engaged to prevent the unionization in this now sensitive scientific facility and how they viewed the connections among union attendance, Communist affiliations, and espionage. The General Leslie Groves wrote to Secretary of War Henry L. Stimson[23]:

1. There has recently been a considerable increase in unionizing attempts among the scientific personnel at the Radiation Laboratory, University of California, Berkeley which is working on the production problem of uranium bombs.

2. The activities of the Federation of Architects, Engineers, Chemists, and Technicians (CIO) Local No. 25 at the Laboratory first came to our attention in connection with the investigation of the transmission of information concerning the DSM Project by a scientist working on the project to espionage agents of the U.S.S.R. It was ascertained by the Military Intelligence Division that [redacted] a scientist working in the Radiation Laboratory had transmitted information to one [redacted] for ultimate transmission to the Soviet Government. It was ascertained that subsequent to the transmission of this information by [redacted] had a clandestine meeting with one Ivanoff of the Soviet Consulate, and later with one Zubilin, Third Secretary of the Soviet Embassy in Washington. [redacted] is a National Committeeman of the Communist Party, U.S.A., and leader of the Alameda County Communist Party. Subsequent investigation indicated that several employees of the Radiation Laboratory are active Communists […] These employees are also members of the FAECT Local No. 25.

3. […] suggested that the most effective and least dangerous method of accomplishing the purpose would be for you to recommend to the President that he call in Mr. Philip Murray, President of the CIO, for a conference, and ask Mr. Murray in the strongest terms to issue directions to the FAECT that all organizational activities with respect to the Radiation Laboratory at the University of California be stopped, and not resumed at any time during the present war.

As a consequence, according to Bird and Sherwin, the Secretary of War Henry L. Stimson wrote to the US President: "Unless this can be at once stopped, I think

---

[22]Rhodes (1986, 488). Mullet (2008, 48).

[23]Mullet (2008, 55). From Brigadier General Leslie R. Groves to the Secretary of War Henry L. Stimson, [with enclosed draft of memorandum to the US President], 17 August 1943. This declassified document was sent by historian Ellen Schrecker to F. David Peat, while he was writing the first Bohm's biography, 2 September 1993, and is deposited at the David Bohm Papers, Folder A.21, Birkbeck College, London. Ellen Schrecker has extensively written about McCarthyism in the US (Schrecker 1986, 1994), in these books Bohm's case is commented upon many times.

the situation is very alarming". Still according to them, "soon afterwards, the CIO was formally asked by the Roosevelt Administration to stop its organizing drive at the Berkeley lab," which eventually was implemented by the head of the Congress of Industrial Organizations (CIO).[24]

Furthermore, the security measures were not limited to the banning of union activities. Restrictions on left wing people at Berkeley reached the team of young physicists around Bohm. Lomanitz had his draft deferment cancelled and was eventually drafted. After an invitation from Oppenheimer to join the atomic project, which would mean to follow to Los Alamos, security officials blocked Bohm's clearance to enlistment on the project on the pretext that he had relatives in Germany. As Bohm recalls, "It was obviously an excuse. I knew it was. Because by then I had become fairly close to these left wing people and I knew that the reason was that they did not like my left wing views". Adding sensitivity to this union activity, there was the belief among security officials that Weinberg had given, on March 29, 1943, "secret information about the project to a local communist official, Steve Nelson, who in turn passed it on to the Soviets". Bohm's perception of the justification for not going to Los Alamos was later corroborated by Bird and Sherwin's analysis: "This was a lie; in fact Bohm was banned from Los Alamos because of his association with Weinberg. He spent the war years working in the Radiation Lab, where he studied the behavior of plasmas". Bohm's scientific work during the war will be commented upon in the next section.[25]

So far, the best documented and studied description of the events surrounding the Berkeley Lab, Oppenheimer's team and suspicion of atomic espionage is the book American Prometheus, written by Kai Bird and Martin Sherwin. While there is documentary evidence of the union activities, its relationship to the Communist Party as well as previous connections of Oppenheimer with Communists and left wing people and causes there is no evidence of espionage by Bohm and "the notion [...] that Oppenheimer could have been recruited as a spy is simply far-fetched". Indeed, still according to Bird and Sherwin, "by the autumn of 1942, it was more or less an open secret around Berkeley that Oppenheimer and his students were exploring the feasibility of a powerful new weapon associated with the atom," which certainly attracted the attention of the Soviets and their allies.[26]

Only in hindsight may we know how sensitive the context was in which Bohm was engaged. It is also in hindsight that we may be sure of the second aspect of Bohm's Marxist commitments. Due to his correspondence and testimonies, we know he remained ideologically committed to Marxism till 1956, when he broke away in the wake of the Soviet invasion of Hungary. He was an avid reader of Marx, Engels,

---

[24]Bird and Sherwin (2005, 176).

[25]Interview of David Bohm by Maurice Wilkins on 1986 July 7, AIP, College Park, MD USA. Bird and Sherwin (2005, pp. 188–193), citation on p. 193; Mullet (2008, 51). We have already seen a reference to Steve Nelson, in this chapter, because he lived for a while in Wilkes-Barre organizing the unemployed workers. He fought in the Spanish Civil War, was a labor organizer, a critic of racial segregation and a member of the Communist Party; in the early 1950s, in Pittsburgh, he was sentenced to jail in the wake of McCarthyism. On Nelson, see Nelson et al. (1981).

[26]Bird and Sherwin (2005), particularly on pp. 187–194.

and Lenin, while at Berkeley, and used to discuss them with Peters, Weinberg, and Lomanitz. References to Marxist beliefs appear scattered through surviving letters from the end of 1951. Still, later, while he was in Brazil, he went on to read Hegel. All the stories of Bohm's engagement with the Communist Party while at Berkeley brought with them heavy consequences after World War II. We will come back to this later.

2.3.2 Bohm's Scientific Training and Work at Berkeley

Bohm's early achievements in physics came out *pari passu* with his own doctoral training at Berkeley. Courses were no longer necessary and Oppenheimer immediately gave him a research problem concerning collision of subatomic particles, more precisely, the scattering of neutrons and protons. While the problem required huge calculation and no deep conceptual analysis, considering his abhorrence for the Caltech courses loaded with calculus, one may wonder why Bohm did not balk at the thought of it. Instead, he worked so intensely that he became exhausted and depressed after completing it and presenting results in a seminar. Oppenheimer liked the results and given the circumstances of the war the work was immediately classified. As Bohm had no clearance to work on military related subjects, an issue surely related to the fact he was being monitored by the security system, an unusual situation arose. He was a doctoral student who could neither analyze his own research results nor present them as a requirement for his Ph.D. degree. The issue was solved with Oppenheimer testifying to the Berkeley administration on the quality of the work and Berkeley granting Bohm a Ph.D. degree without the presentation of a doctoral dissertation. As Bohm recalled, years later, "I did manage to finish my Ph.D. degree doing a more restrictive problem that Oppenheimer had suggested. The scattering of neutrons and protons which involved a fair amount of numerical calculations. But I finished it and he apparently found some use for it at Los Alamos. It was branded secret. But I got my degree out of it anyway on his word in 1943".[27]

Bohm was then hired by the Radiation Laboratory to work on the equipment derived from the new cyclotron accelerator, that is, the calutron. This work, of most important use, for military purposes, was related to its use to focus ion beams to separate Uranium-235 isotopes. In nature uranium is found mainly in the form of its 238 isotope which is not fissile. Thus if nuclear fission was to be used as a source for nuclear energy, including as weapons, the use of uranium requires the relative enrichment of the amount of its 235 isotope when compared to the 238 isotope. While the military interest in the problem was clear cut, it required some previous basic knowledge and Bohm was engaged in this task. Reminiscences of Robert Serber, the theoretical physicist in charge of introducing the physicists arriving at Los Alamos

---

[27]Interview of David Bohm by Maurice Wilkins on 1986 July 7, AIP, College Park, MD USA.

on the basics of nuclear physics to be used in the building of the bombs, said the following about how this task was undertaken[28]:

> Oppy had assembled a small theoretical group. Lawrence had a project going up to the hill to devise a way to separate U^{235} from U^{238} electromagnetically—the project that developed the calutron electromagnetic separators that the Manhattan Project eventually operated at Oak Ridge. This bunch of kids Oppenheimer had put together, half a dozen or so, were working on calculating orbits in the magnetic field and that kind of thing.

Ion beams of uranium isotopes may be separated passing them through the magnetic and electric fields of the cyclotron. To obtain ion beams from metallic uranium this metal needs to be at very high temperatures and ion gases at very high temperatures were the state of matter physicists were calling plasma, a kind of a fourth state of matter to distinguish it from the known solid, liquid and gaseous states. However, the understanding of the physical properties and behavior of such a state was scant and this was the road which lead Bohm to the study of plasma. Initially he worked with Stanley Frankel and Alfred Nelson but they left for Los Alamos. Bohm kept at it but without anybody to talk and the work advanced at a slow pace. Then, in the winter of 1943 a British team arrived at Berkeley as part of the joint war effort. The team included Harrie Stewart Wilson Massey, Eric Henry Stoneley Burhop, Maurice Hugh Frederick Wilkins, and Mark Oliphant who was the leader of the team and had previous experience in the use of accelerators. Massey, Burhop, and Oliphant were Bohm's senior by about ten years while Wilkins was the same age as Bohm (Fig. 2.12).

Bohm would maintain close connections with this team, particularly with Wilkins, who would move from physics to molecular biology and become the "third man" in the trio awarded the Nobel Prize for the discovery of the structure of DNA. Wilkins shared with Bohm interests in philosophy. Later on, he offered Bohm a chapter in a Festschrift in honor of Bohm's 70th birthday.[29]

After the war, Bohm presented some qualitative results of his work with plasma at the 273rd meeting of the American Physical Society which was held at the University of California, Berkeley, on 12–13 July, 1946. For Bohm it was a great experience as it was the first large scientific meeting he had ever attended. According to R. T. Birge, APS local secretary, "The attendance was by far the largest at any meeting on the Pacific Coast in the history of the Society," gathering around 500 physicists. However, the qualitative nature of Bohm's scientific communication did not appeal much to the audience. Only later, with the declassification of some of these war studies, the significance of Bohm's work would be revealed. Indeed, after the World War II the US government slowly began to declassify a certain number of results from the atomic project which were no longer considered sensitive. Thus in 1949 in the volume "The Characteristics of Electrical Discharges in Magnetic Fields," edited by A. Guthrie and R. K. Wakerling, the main results of this team were gathered and published. The extent of Bohm's work may be seen from the fact that five out

---

[28] Serber (1992, xxix).

[29] Wilkins (2003), (1987). Interview of David Bohm by Maurice Wilkins, 1986–1987, Niels Bohr Library and Archives, American Institute of Physics, College Park, MD USA.

Fig. 2.12 Robert
Oppenheimer, Bohm's
doctoral supervisor, and
Ernest Lawrence at the
184-Inch Cyclotron at
Lawrence's Radiation
Laboratory, Berkeley. This
accelerator was completed in
1946. *Credits* Digital Photo
Archive, Department of
Energy (DOE), courtesy of
AIP Emilio Segrè Visual
Archives

of the eleven papers were authored or co-authored by him. Furthermore, what it is
known as Bohm diffusion, was published there. This is a coefficient measuring the
diffusion of electrons through directions which are transverse to the magnetic field:
$D_\perp \approx \frac{10^5}{16H}\left(\frac{kT_e}{e}\right)$, "where H is in thousands of gauss and kT_e/e is in volts".[30]

The authors state that "the exact value of $D_\perp$ is uncertain within a factor of 2 or
3". The most important, however, was the conclusion that "the rate of drain diffusion
varies as 1/H, in contrast to that of collision diffusion, which varies as $1/H^2$". Drain
diffusion was the term used to describe the model conceived by Bohm to describe
the combination of the individual movement of the ions, subjected to instabilities
and fluctuations, which produced a collective oscillation of all ions leading to the
appearance of fields and then the "drain" of the ions. The predicted phenomenon

---

[30]Bohm's communication to the APS meeting is Bohm (1946a), Birge's comment on the meeting
are in Physical Review, 70(5–6), p. 443, 1946. Bohm's works in the published volume are (Bohm
1949a, b, c); (Bohm et al. 1949a, b). The diffusion equation is in Bohm et al. (1949a, 65). The
published volume is Guthrie and Wakerling (1949).

would avoid the ion beam from being tightly focused. The previous models, named "collision diffusion" only considered changes in the ion movements deriving from random collisions among them. For all practical purposes, the implication of this new model was that with it diffusion is "much more rapid" than in the previous model. If this previous model had been right, small increases in the magnetic field confining the ions, that is the plasma, would lead to higher confinement periods of time. Bohm's model had far-reaching implications, if right, smaller confinement times would imply the practical impossibility of magnetic confined fusion. When the quest for the civil use of nuclear fusion was launched, after the war, for a while Bohm's result seemed a negative result indicating constraints in this direction. Only years later, with the development of the tokamak, was it realized that Bohm's model could be overcome by other models. According to the historians of physics Lilian Hoddeson and colleagues, Bohm suspected instabilities were causing turbulence in the plasma and "worked out intuitively, essentially on dimensional grounds, a description of the phenomenon now known as 'Bohm diffusion;'" and they added "the microscopic theory for this phenomenon would not be worked out for two more decades".[31]

Bohm's achievements were thus no small deeds for a recent Ph.D., a postdoc we would call today. This is the kind of work that cannot evaluated using data from citations of papers as his findings were published as book chapters thus are not recorded on data bases such as the *Web of Science*. Indeed, in this database, these papers do not appear on the list of Bohm's publications but the term "Bohm diffusion" appeared in 339 titles of papers till August 2018; the first time it was used in a title was in 1963 but its widespread use began in early 1990s.[32] For Bohm, these works on plasma were also opportunities to meet those British physicists and for Bohm getting professional recognition among them. In fact, as we will see later, the respect he gained from physicists such as Burhop and Wilkins would help him in the future battles over the reinterpretation of quantum mechanics.

Bohm's work on plasma coalesced a scientific style to be further developed throughout his life. He was inspired by daring analogies, often drawing together meanings from distant domains of experience. In addition, he shared a realist view which requires the imagination of pictures of the systems under study. Furthermore, of course, all these analogies and pictures should be submitted to the scrutiny of and developed through the mathematical language. In the concrete case, as we have already seen, he imagined all the ions collectively oscillating and this oscillation creating fields which interfere with the movement of each single ion. This imagination was inspired, or supported, by the analogy he developed with the social thought he was toying with at the time. Indeed, he was concerned at that time with the interference of collective, social, movements, with singular, free movements of individuals. For him it was kind of a trade-off between his former thought about the role of individualism in American society and the new way of thinking in terms of socialism to be reached through collective movements. As he recalled years later, "The

---

[31]Hoddeson et al. (1992, 535).

[32]*Web of Science*, consulted on August 8, 2018.

plasma became very interesting to me. I could see that this was a kind of analogy to the problem of the individual and the society. You had in the plasma what I called collective behavior, that is, oscillations. Every plasma can oscillate. When all the electrons move together, they produce an electric field that draws them back so that they'll oscillate. They oscillate in a coherent way which belongs to the whole. I call that a collective movement". Still, "How was this collective movement maintained in spite of the random basis of the electrons? You see, this was the kind of interesting social question. It was rather like society, everybody moving in his own way and you have certain social, collective tendencies still exist". In a book published around the time of these recollections, Bohm wrote along similar lines[33]:

> My insights sprang from the perception that the plasma is a highly organized system which behaves as a whole. Indeed in some respects, it's almost like a living being. I was fascinated with the question of how such organized collective behavior could go along with the almost complete freedom of movement of the individual electrons. I saw in this an analogy to what society could be, and perhaps as to how living beings are organized.

The analogies Bohm used to develop his physical ideas led the historian of science Alexei Kojevnikov (2002) to argue that from this early work on plasma to his later works with David Pines and Eugene Gross on the use of collective variables for the study of plasmas and metals, Bohm's productive scientific work was driven by political metaphors. Kojevnikov arrived at these conclusions in Bohm's case following wider research of philosophical assumptions, or motivations, among condensed matter physics, a domain usually considered wanting of philosophical considerations. Kojvenikov had studied cases of Soviet physicists who dealt with issues such as individual movements of systems and their collective properties, in other words the freedom of individuals and the constraint of their collective behavior. With Bohm's case, Kojevnikov found similarities in the use of political and philosophical metaphors across the two camps, USSR and the US. Thus, he argued, "collectivism, provided intuitions for a group of condensed matter theorists, who tried to solve the basic problem of freedom for atoms and electrons along collectivist lines. Initially, only a minority of physicists, most of them socialists, designed physical and mathematical models for the collective behavior of microscopic particles". Yet Kojevnikov was cautious to warn about the peculiarities of each case he was studying: "Their models differed according to the specific physical problem and the various meanings of "collectivism" in socialist thought". In the same direction, while in a less construed manner, the writer F. David Peat argued that "one description (collective coordinates) deals with the collective vibrations, while the other (individual coordinates) explains free individual movement. Yet because the two descriptions are part of a single whole, the collective motion of the whole is enfolded within the random, individual movement, and vice versa".[34] It should be noted, finally, that political metaphors, philosophical views, and physics problems were all part of the scientific reasoning of a creativity mind as Bohm's.

---

[33] Interview of David Bohm by Maurice Wilkins on 1986 July 7, AIP, College Park, MD, USA. The citation is from Bohm and Peat (1987, 5).

[34] Kojevnikov (2002, 162). Peat (1997, 67).

During the war, research of military interest was not Bohm's only duty. After Oppenheimer left Berkeley for Los Alamos, he and Weinberg assumed full responsibility for the quantum mechanics course. Bohm enjoyed it and this may have contributed to his ongoing passion with the understanding of the quantum. He used to discuss with Weinberg Bohr's complementarity. Weinberg looked for compatibility between Marxist philosophy and the complementarity view, which he considered something dialectical. As for Bohm, "at that time, I was convinced that Bohr's approach was the right approach and for many years I continued with Bohr's approach". After the plasma work and at the end of the war, Bohm went to work on different subjects at Berkeley, including the building of synchrocyclotron machines, with Robert Serber, Edwin M. McMillan, and Leslie L. Foldy. The latter had gone to Berkeley to do his Ph.D. under the supervision of Oppenheimer but worked with Bohm for a while on the theory of synchroton accelerators. Bohm resumed work with Oppenheimer, this time on superconductivity, but apparently nothing was achieved; while later, at Princeton, he would go back to the subject (Bohm 1949d). In addition to these different physics subjects, he spent his spare time working alone on quantum fields. He worked on the infinities appearing in the quantum field theories and had the experience of having a paper with this work refused by the journal *Physical review* but never went back to the subject. He also worked on a new theory the British physicist Paul Dirac had put forward in 1942 to quantize the electromagnetic field. He recalls that Oppenheimer discouraged him from further pursuing the subject as Dirac himself had abandoned it. Bohm did not concede and presented the work at the American Physical Society (APS) held on September, 19–21, 1946 in New York.[35]

Bohm thought it was at the APS meeting that his work caught the attention of John Archibald Wheeler, from Princeton, who invited him to join this university. Wheeler's reminiscences, however, do not make reference to this. Indeed, as he wrote later in his autobiography, "I was largely responsible for bringing Bohm to Princeton in the first place. Shortly after World War II, I had visited Berkeley and, at my department's request, interviewed Bohm. Upon my favorable recommendation, Princeton offered him a temporary appointment, and he joined the department in 1947". The reference to the "department's request" suggests there was wider interest at Princeton in attracting Bohm. Indeed, either through his work during the war or the favorable opinions of both Oppenheimer and the British team, there is no doubt he was a good candidate to occupy a place in the flourishing field of physics in post war America. In early 1947 David Bohm took up his post at the Physics Department at Princeton.[36]

---

[35]Interview of David Bohm by Maurice Wilkins on 1986 July 7, AIP, College Park, MD, USA. Bohm and Foldy (1946), (1947). Bohm (1946b).

[36]Wheeler and Ford (1998, 216).

References

Andrade, V.F.P.: Vendedores de estrelas—A existência de outras galáxias pela mídia de massa norte-americana na década de 20. Universidade Federal da Bahia, Brazil, Ph.D. Dissertation (2017)

Bird, K., Sherwin, M.J.: American Prometheus—The Triumph and the Tragedy of J. Robert Oppenheimer. Alfred Knopf, New York (2005)

Black, J.M., Fielding, J., Morin, B. (eds).: Anthracite Photographers: Photographers of Anthracite, [Catalogue of exposition], Anthracite Heritage Museum, Scranton, PA (2018)

Bohm, D.: Excitations of plasma oscillations. Phys. Rev. **70**(5–6), 448 (1946a)

Bohm, D.: Relations of Dirac's new method of field quantization to older theories. Phys. Rev. **70**(9–10), 795 (1946b)

Bohm, D.: Qualitative description of the arc plasma in a magnetic field. In: Guthrie, A., Wakerling, RK, McGraw-Hill, NY, (pp. 1–12) (1949a)

Bohm, D.: Minimum ionic kinetic energy for a stable sheath. In: Guthrie, A., Wakerling, R.K. (pp. 77–86), (1949b)

Bohm, D.: Theoretical considerations regarding minimum pressure for stable arc operation. In: Guthrie, A., Wakerling, R.K. (pp. 87–106), (1949c)

Bohm, D.: Note on a theorem of Bloch concerning possible causes of superconductivity. Phys. Rev. **75**(3), 502–504 (1949)

Bohm, D, Burhop, E.H.S., Massey, H.S.W.: The use of probes for plasma exploration in strong magnetic fields. In: Guthrie, A., Wakerling, R.K. (pp. 13–76) (1949)

Bohm, D., Burhop, E.H.S., Massey, H.S.W., Williams, R.W.: A study of the arc plasma. In: Guthrie, A., Wakerling, R.K. (pp. 173–333) (1949)

Bohm, D., Foldy, L.: The theory of the synchrotron. Phys. Rev. **70**(5–6), 249–258 (1946)

Bohm, D., Foldy, L.: Theory of the Synchro-Cyclotron. Phys. Rev. **72**(8), 649–661 (1947)

Bohm, D., Peat, F.D.: Science, Order & Creativity. Routledge, London (1987)

Eddington, A.S.: Relativity Theory of Protons and Electrons. The Macmillan Company, London (1936)

GARCHIVE 1935: [Published by the Senior Class of G.A.R. Memorial High School], Wilkes-Barre, Pennsylvania (1935)

Gernsback, H.: Interplanetary travel. Amazing Stories **1**(11), 981 (1927)

Guthrie, A., Wakerling, R.K.: The Characteristics of Electrical Discharges in Magnetic Fields. McGraw-Hill, New York (1949)

Hobsbawm, E.: Age of Extremes—The Short Twentieth Century 1914–1991. Abacus, London (1995)

Hobsbawm, E.: How to Change the World: Tales of Marx and Marxism. Little Brown, London, London (2011)

Hoddeson, L., Schubert, H., Heims, S.J., Baym, G.: Collective phenomena. In: Hoddeson, L., Braun, E., Teichmann, J., Weart, S. (eds.) Out of the Crystal Maze—Chapters from the History of Solid-State Physics, pp. 489–616. Oxford University Press, New York (1992)

Kragh, H.: Quantum Generations—A History of Physics in the Twentieth Century. Princeton University Press, Princeton, NJ (1999)

Kragh, H.: Higher Speculations—Grand Theories and Failed Revolutions in Physics and Cosmology. Oxford University Press, Oxford (2011)

Kojevnikov, A.: David Bohm and collective movement. Hist. Stud. Phys. Biol. Sci. **33**, 161–192 (2002)

Mullet, S.K.: Little man: four junior physicists and the red scare experience. Ph.D. Dissertation, Harvard University (2008)

Nelson, S., Barrett, J.R., Ruck, R.: Steve Nelson: American Radical. University of Pittsburgh Press, Pittsburgh, PA (1981)

Paty, M.: L'analyse critique des sciences, ou le tétraèdre épistémologique. L'Harmattan, Paris (1990)

Paty, M.: Einstein philosophe. Presses Universitaires de France, Paris (1993)

Peat, F.: David. Infinite Potential: The Life and Times of David Bohm. Addison Wesley, Reading, Ma (1997)

Rhodes, R.: The Making of the Atomic Bomb. Simon & Schuster, New York (1986)

Schrecker, E.: No Ivory Tower: McCarthyism and the Universities. Oxford University Press, New York (1986)

Schrecker, E.: The Age of McCarthyism: A Brief History with Documents. Bedford Books of St. Martin Press, Boston (1994)

Schweber, S.S.: The empiricist temper regnant—theoretical physics in the United-States 1920–1950. Hist. Stud. Phys. Biol. Sci. **17**, 55–98 (1986)

Schweber, S.S.: In the Shadow of the Bomb: Bethe, Oppenheimer, and the Moral Responsibility of the Scientist. Princeton University Press, Princeton, NJ (2000)

Serber, R.: The Los Alamos primer: the first lectures on how to build an atomic bomb [Annotated by Robert Serber; edited with an introduction by Richard Rhodes]. University of California Press, Berkeley (1992)

Seth, S.: Crafting the Quantum—Arnold Sommerfeld and the Practice of Theory, 1890–1926. The MIT Press, Cambridge, Ma (2010)

Smythe, W.R.: Static and dynamic electricity. McGraw-Hill, New York (1939)

Wheeler, J.A., Ford, K.: Geons, Black Holes, and Quantum Foam. Norton, New York (1998)

Whittaker, E.T., Watson, G.N.: A Course of Modern Analysis—An Introduction to the General Theory of Infinite Processes and of Analytic Functions: With an Account of the Principal Transcendental Functions. [First edition 1902]. University Press & Macmillan Co., Cambridge and New York (1943)

Wilkins, M.H.F.: Complementarity and the union of opposites. In: Hiley, B.J., David Peat, F. (eds). Quantum Implications—Essays in honour of David Bohm, pp. 338–360, Routledge and Kegan Paul, London (1987)

Wilkins, M.H.F.: The Third Man of the Double Helix: The Autobiography of Maurice Wilkins. Oxford University Press, Oxford (2003)

Wolensky, R.P., Hastie Sr., W.A.: Anthracite Labor Wars: Tenancy, Italians, and Organized Crime in the Northeastern Pennsylvania 1897–1959. Canal History and Technology Press, Easton, PA (2013)

Chapter 3
Teaching and Doing Research at Princeton, Caught up in the Cold War Storms (1947–1951)

The years Bohm spent in California shaped his scientific style enabling him to exhibit his potential skills as a physicist and his political and social mind. In the almost five years he stayed at Princeton he matured as a creative scientist but also experienced the most traumatic and damaging events. He pursued a research program dedicated to the understanding of plasma, worked out the collective variables approach, and extended these ideas to the study of metals. He attracted talented graduate students such as Eugene Gross and David Pines and built up an entirely original interpretation of quantum theory, the so-called hidden variables interpretation or, still, causal interpretation. He came of age as a skilled science writer producing a textbook on quantum physics still in use today. And yet, he was a victim of anti-communist campaigns which took hold in American society at the apex of the Cold War. He felt coerced into leaving the United States in order to further pursue his scientific and professional life. In this chapter we will successively see his pedagogical work and research on plasma and metals, and troubles on the American political scene. Then we spend time analyzing his suggestion for a new interpretation of quantum mechanics and the outcome of the political situation.

3.1 Teaching Quantum Mechanics at Princeton

At Princeton Bohm began teaching undergraduate courses but quickly moved to teach quantum mechanics to graduate students. It would be a life changing experience as the preparation of these courses constituted the raw material of what would become his first book, published in 1951. We do not know much about how Bohm prepared the book, only that he brought his Bohrian inclinations to Princeton as far as the interpretation of quantum mechanics is concerned, which was a result of conversations with Joseph Weinberg and his mix of complementary and dialectic and from the overall intellectual ambiance around Oppenheimer at Berkeley. And, of course, we know the unique features of this textbook and how was it received by Bohm's fellow

© Springer Nature Switzerland AG 2019
O. Freire Junior, *David Bohm*, Springer Biographies,
https://doi.org/10.1007/978-3-030-22715-9_3

physicists. Before delving into the contents of the book, the clarity of the text is worth mentioning, a characteristic also present in all four of his other scientific books he would later write and in the many popular and more philosophically inclined books he would author or co-author throughout his life. For one of the early reviewers, "Dr. Bohm's book is a rare example of expressive, clear scientific writing," while for another, "here is a well-written textbook on quantum mechanics which introduces the student to the subject with primary emphasis on the clarity of exposition of the fundamental concepts."[1]

What was remarkable about Bohm's *Quantum Theory* compared to the available textbooks were its ontological realistic assumption, the treatment of measurement, the analysis of the Einstein-Podolsky-Rosen argument (EPR), plus a consideration that Niels Bohr was right in rebuking such an argument, and the new mathematical presentation of this argument, which included the two-levels states for the spin state of a pair of electrons. Shared with other textbooks was his defense of the standard interpretation of quantum mechanics, which led him to present and support Niels Bohr's views on the principle of complementarity as well as to consider Bohr's stances in the debates with Einstein on the EPR argument favorably. Due to the importance of this book for Bohm's later views on quantum mechanics and his permanent interest in understanding the quantum, it is worth examining in some detail the content of this book.

Bohm dedicated a full chapter to "An attempt to build a physical picture of the quantum nature of matter." In this chapter he introduced the Bohrian "principle of complementarity" to his readers. In addition, the notion of "potentiality" is introduced by Bohm with independence and antecedence of Werner Heisenberg's idea of *potentia*.[2] Bohm defined it as a new concept which considers the quantum properties of matter as potentialities. For him "this new concept considers these properties as incompletely defined potentialities, the development of which depends on the systems with which the object interacts, as well as on the object itself." Speaking about momentum and the position of an electron, Bohm says they may be called "interwoven variables" but this would be an inadequate description "since it does not include the idea that the very existence of either requires a certain degree of indefiniteness of the other." Then he suggests the term "interwoven potentialities" as it represents "opposing properties that can be comparatively well defined under different conditions." Potentialities are thus linked to the principle of complementarity, as he writes in a nutshell[3]:

> Thus, an individual electron must be regarded as being in a state where these variables are actually not well defined but exist only as opposing potentialities. These potentialities

---

[1] Bohm (1951), Bohm's books are Bohm (1957, 1965, 1980) and Bohm and Hiley (1993). The reviews are Corson (1952) and Inglis (1952), respectively.

[2] Bohm (1951, 158–161). "This probability concept is closely related to the concept of possibility, the 'potentia' of the natural philosophy of the ancients such as Aristotle; it is, to a certain extent, a transformation of the old 'potentia' concept from a qualitative to a quantitative idea" (Heisenberg 1955, 13).

[3] Bohm (1951, 132 and 159).

complement each other, since each is necessary in a complete description of the physical processes through which the electron manifests itself; hence, the name "principle of complementarity."

Nevertheless, Bohm's identification of potentialities with complementarity was not without problems. For the complementarity view preached by Niels Bohr waves and particles were pictures, representations, embedded with concepts from the classical theories of physics. Due to the limitations introduced by quantum physics on the classical concepts, they were to be used in a mutually exclusive manner according to the experimental setups. Thus waves and particles had no ontological implications. However, for David Bohm waves and particles were real physical phenomena, in fact, quantum properties of matter, furthermore, parts of physical reality, thus ontological entities. In the textbook Bohm was apparently not aware of these nuances.

Next, Bohm presented the debate between Bohr and Einstein, through the presentation of EPR argument and Bohr's answer, concluding that quantum mechanics corroborated Bohr's views and not those shared by Einstein. Bohm's analysis of EPR, which he dubbed ERP inverting the order of the names of Boris Podolsky and Nathan Rosen, is noticeably sophisticated and modern. He rightly identifies the assumption in EPR's thoughts which run counter to quantum mechanics, namely, "that the analysis given by ERP involves in an integral way the implicit assumptions [...] that the world is actually made up of separately existing and precisely defined 'elements of reality'." Later on, philosophers of science would call "separability" this assumption in Einstein's thoughts.[4] For Bohm, it was precisely this assumption which is not corroborated by quantum mechanics itself.

Furthermore, to arrive at this conclusion he presented the EPR argument, or to be more precise, the experiment Einstein had idealized to exhibit the point, in novel mathematical terms. Instead of describing the two distant particles through their position and momentum coordinates, which were continuous coordinates, Bohm thought of a couple of electrons having interacted and moving far away from each other. Now Bohm presented them through their spin projections, whose quantum state is the "singlet" state for a zero total angular moment. These are discrete coordinates, instead of continuous, with only two values. The far-reaching implications of the mathematical simplifications would be very useful years later in the wake of Bell's 1965 theorem and its experimental tests. Thus for Bohm the pair of entangled particles would have the quantum state of their components of spin, with total spin zero, written as:

$$\psi_0 = \frac{1}{\sqrt{2}}(\psi_c - \psi_d)$$

where $\psi_c = u_+(1)u_-(2)$ and $\psi_d = u_-(1)u_+(2)$; and u_+ and u_-, are "the one-particle spin wave functions representing, respectively, a spin $\hbar/2$ and $-\hbar/2$, and the argument (1) or (2) refers, respectively, to the particle which has this spin." Thus,

---

[4]Bohm (1951, 615). See Howard (1985, 2017).

$$\psi_0 = \frac{1}{\sqrt{2}}[u_+(1)u_-(2) - u_-(1)u_+(2)]$$

Nowadays this is the technical expression of the quantum entanglement of a pair of quantum systems with zero total spin. Returning to Bohm, "because the wave function has definite phase relations between ψ_c and ψ_d, the system must cover the states corresponding to ψ_c and ψ_d simultaneously." Thus, Bohm concludes, "for a given atom, *no* component of the spin of a given variable exists with a precisely defined value, until interaction with a suitable system, such as a measuring apparatus, has taken place."[5]

For measurements, Bohm did not adopt the treatment John von Neumann had presented in his canonical book. For von Neumann, measurement should be described through different presumptions from those ruling the time evolution of quantum systems, such as described by Schrödinger equations. Von Neumann suggested quantum measurement should be treated as a distinct evolution, instantaneous, ruled by the mathematical operators he called operator projections.[6] For Bohm, however, measurement should be described by the same physical and mathematical device as Schrödinger equation and he modelled measurements in quantum mechanics appealing to the coupling of the quantum states from both the system under study and the measurement apparatus, dealing with the phases of these states, and appealing to the random nature of these phase factors.

The presentation of the quantum theory of the measurement process was another instance where Bohm's thoughts were close but not identical to those of Niels Bohr. Bohm stated "if the quantum theory is to be able to provide a complete description of everything that can happen in the world, however, it should be able to describe the process of observation itself in terms of the wave functions of the observing apparatus and those of the system under observation;" which would not be accepted by Bohr without further and restrictive clarifications. Bohm was aware of the paradox—Schrödinger's cat—arising from using quantum mechanics to describe measurement devices, as he wrote "if it were necessary to give *all* parts of the world a completely quantum-mechanical description, a person trying to apply quantum theory to the process of observation would be faced with an insoluble paradox." This was because "he would then have to regard himself as something connected inseparably with the rest of the world." He went on, "on the other hand, the very idea of making an observation implies that what is observed is totally distinct from the person observing it." How did he solve this paradox? Through a very Bohrian argument, the role of classical description on the description of measurements, Bohm states: "this paradox is avoided by taking note of the fact that all real observations are, in their last stages, classically describable."[7] The book was thus an exhibition of independent thinking as Bohm presented quantum theory without full alignment with Bohr, Einstein, and

---

[5]Bohm (1951, 616–621).

[6]Bohm (1951, 583) warned readers that his presentation was different from von Neumann's: "For another treatment of this problem, see von Neumann (1932)."

[7]Bohm (1951, 583–585). On the measurement problem, see also my *The Quantum Dissidents* (Freire Junior 2015), particularly on pp. 125–128 and 141–174.

von Neumann. The three were giants of physics at the time, all three around him at Princeton; the first through the influence of Oppenheimer and the latter two in person at the Institute for Advanced Study at Princeton.

The book presented arguments against the introduction of hidden variables in quantum mechanics—if they were introduced, quantum mechanics predictions would not be all true—a stance he would revise in the coming years and wrote about the "indivisible unity of the world" and about "possible reason for analogies between thought and quantum processes," a subject he would ponder and write about through-out his entire life.[8]

The book was well received, praised through reviews in professional journals such as *Physics Today*, *Nature* and *American Journal of Physics*, and was reprinted a few years later. It has since been reprinted several times including an inexpensive Dover books edition.[9] However, when the book was released Bohm was in the midst of the political turmoil related to American anti-communist witch-hunt. He was working on his new interpretation of quantum mechanics, which led him away from the standard presentations of quantum mechanics, two subjects we will address later.

3.2 Research Into Plasma Goes on

At Princeton David Bohm began to look for a new research program. While tuned to nuclear physics and quantum field theory, he balked at mainstream physics and went to develop further the study of plasma. His attention was caught at the Pocono conference, held on 30 March–2 April 1948. This was the second in a series of three conferences, the first one at Shelter Island, which had been held on 2–4 June 1947, while the third one was the Oldstone conference, held on 11–14 April 1949. They aimed to bring together the most promising American theoretical physicists with some of their seniors to discuss fundamental issues in physics which had been put on a back burner since the engagement in the war effort. Bohm's invitation to these conferences is in itself evidence of the expectations American physicists had regarding his work. The Pocono conference was marked by two presentations both concerning the renormalization issue in quantum electrodynamics. The first one by Julian Schwinger was a day-long talk and second one was by Richard Feynman who presented his newly developed diagrams. After the conference and the circulation of the work of the Japanese physicist Sin-Itiro Tomonaga on the same subject, with sev-eral distinct contributions including those by the English-born Freeman Dyson, the equivalence among these approaches was eventually set and their works became the canonical treatments for the infinities which had plagued quantum electrodynamics for decades. All this work was rewarded later with the 1965 Nobel Prize to Feynman, Schwinger, and Tomonaga. This story provides some background to Bohm's choice

---

[8] Bohm (1951, 622 and 161–170), respectively.

[9] Corson (1952), Inglis (1952), and Le Couteur (1953). The Dover edition, unabridged and unaltered, was published in 1989.

for plasma work. According to historians Hoddeson, Schubert, Heims, and Baym, it was at the Pocono conference, listening to the long lecture by Schwinger, that Bohm had the idea to use Schwinger's canonical transformation approach as the way to mathematically treat the collective coordinates he was dealing with in his plasma work. Indeed, neither at Shelter Island nor at Pocono was Bohm attracted to work on the treatment of infinities in quantum electrodynamics and the conflict between its predictions and the recently discovered experimental results known as Lamb shift.[10]

Bohm's choice of a research program deserves more reflection. According to the historian of physics Alexei Kojevnikov, "at Princeton Bohm concentrated increasingly on plasma as his own, independent research subject. Plasma was not considered a prestigious, fundamental topic, in part because it did not seem to have a military importance." Kojevnikov went on to say that "war-time studies of plasma had not helped the Manhattan Project; their military relevance became clear only a few years later, with the start of the H-bomb race," to state that "in Bohm's eyes, however, plasma was more fundamental than atomic nuclei, in particular from the philosophical point of view." Kojevnikov's conclusion is supported by Bohm's own later recollections: "First of all, it was a sort of autonomous medium; it determined its own conditions, it had its own movements which were self-determined, and it had the effect that you had collective movement, but all the individuals would contribute to the collective and at the same time have their own autonomy."[11]

David Bohm developed this research program mostly with two graduate students, Eugene Gross and David Pines. Gross met Bohm at Princeton and has a vivid memory of their first meeting. Bohm gave a seminar on plasma and Gross was struck by the research program Bohm had presented, took notes and then presented them to Bohm, after which Bohm took him on as a doctoral student. According to Gross' recollections, this was the vast panorama presented by Bohm at his talk[12]:

> The first part dealt with the plasma as a distinct state of matter, with an organization different from the solid, liquid, and gas. The charge screening and lack of velocity locking was emphasized. The second part had to do with the widespread occurrence in discharge tube physics, in astrophysics and in chemistry. […] The third part dealt with metals viewed as quantum plasmas. The main tool of analysis was the linearization of the equations of motion for products of creation and annihilation operators by means of the random phase approximation.

Bohm's other doctoral student at Princeton was David Pines. Together they extended the approach earlier intended for plasma to the study of metals as a dense gas. Pines would later extend this approach to superconducting metals, nuclear physics, superfluidity, quantum liquids, and other topics. In addition to several specific results they developed a method which has since been highly influential in

---

[10]Hoddeson et al. (1992, 536). On the conferences and the debates on electrodynamics, see Schweber's chapter "Three conferences: Shelter Island, Pocono, and Oldstone," in Schweber (1994, 156–205) and Mehra (1994, 223–279).

[11]Kojevnikov (2002, 170) cites Bohm from an interview by Lillian Hoddeson, held in 1981 and nowadays available at the American Institute of Physics.

[12]Gross (1987, 46).

physics subjects. According to Pines, the method he and Bohm developed "both in [their] studies of the equations of motion of density fluctuations and in [their] collective description of plasma waves plus single particles, [was] called at first the plasma approximation;" a method they would later call the "random phase approximation." (Figs. 3.1, 3.2 and 3.3).[13]

The research with Gross and Pines bore fruit in the long term. Five of the papers resulting from this research are now among the ten papers Bohm authored or co-authored and have now more than 300 citations. As well as these, there is also a paper by Pines alone and an extension of this approach, mainly by Pines, to superconducting metals, with John Bardeen, to a comparison between nuclear spectra and superconducting metals, with Aage Bohr and Ben Mottelson, and to quantum liquids, with Philippe Nozières. The influence of this approach, far beyond plasma and metals, was testified by Ben Mottelson who acknowledged the influence of this approach in his Nobel Prize speech in 1975. In their work with Aage Bohr models

Fig. 3.1 David Pines was among the first doctoral students of David Bohm. (L–R): David Pines, Christopher Pethick, Lev Pitaevski, Valery Shikin, Segre Anisimov and William McMillan. October 24–25, 1976. *Credits* Department of Physics, University of Illinois at Urbana-Champaign, courtesy of AIP Emilio Segrè Visual Archives

---

[13]Pines (1987, 78).

Fig. 3.2 David Bohm was among the most promising young American theoretical physicists to attend the Shelter Island conference, June 1947. Shelter Island conference participants pose for photograph. L–R are: Rabi, I. I.; Pauling, Linus Carl; Van Vleck, John Hasbrouck; Lamb, Willis Eugene; Breit, Gregory; MacInnes, Duncan Arthur; Darrow, Karl Kelchner; Uhlenbeck, George Eugene; Schwinger, Julian Seymour; Teller, Edward; Rossi, Bruno Benedetto; Nordsieck, Arnold Theodore; Von Neumann, John; Wheeler, John Archibald; Bethe, Hans Albrecht; Serber, Robert; Marshak, Robert Eugene; Pais, Abraham; Oppenheimer, J. Robert; Bohm, David; Feynman, Richard Phillips; Weisskopf, Victor Frederick; Feshbach, Herman. *Credits* Fermilab Photograph, courtesy AIP Emilio Segrè Visual Archives, Marshak Collection

of nuclei, "it was a fortunate circumstance for us that David Pines spent a period of several months in Copenhagen in the summer of 1957, during which he introduced us to the exciting new developments in the theory of superconductivity."[14]

---

[14]Bohm's papers with Pines or Gross with more than 300 citations are Bohm and Gross (1949a, b), Pines and Bohm (1952), Bohm and Pines (1951, 1953), Pines (1953), Bardeen and Pines (1955), Bohr et al. (1958), Nozières and Pines (1990), Ben Mottelson, The Nobel Lecture, available at: https://www.nobelprize.org/nobel_prizes/physics/laureates/1975/mottelson-lecture.html.

Fig. 3.3 (L–R): John
Wheeler, Eugene Wigner.
Wheeler was the physicist
responsible for bringing
Bohm to Princeton
University. *Credits* AIP
Emilio Segre Visual
Archives, Wheeler
Collection

3.3 Trapped in the Cold War Storms

Just when research into plasma and metals and the writing of the *Quantum Theory*
book were flourishing, Bohm's promising career in American physics collapsed in the
Cold War turmoil that took hold in the US. In early 1949 he was called to the House
Un-American Activities Committee (HUAC), a committee of the United States House
of Representatives. It was investigating atomic espionage and communist activities at
the Berkeley Radiation Laboratory. According to the press coverage, "the committee
is investigating charges that a Red spy operated at the Berkeley laboratory during the
war. It contends secret information on atomic developments were given to Arthur
Alexandrovich Adams, described as a 'super spy' who fled to Russia in 1945."[15]
They asked Bohm if he was a Communist. Bohm appealed to the Constitution Fifth

---

[15]"A-Bomb Scientist Evades Red Query": "David Bohm, a Princeton university Professor who
helped develop the atom bomb, yesterday refused to say whether he is a Communist on grounds it
might incriminate him," *The Washington Post*, 26 May 1949. "Congressional spy hunters questioned
an alleged Communist espionage agent and two atomic scientists today without producing evidence
of any deals between them," *The New York Times*, 27 April 1949. For the "Scientist X", "NAMED
'SCIENTIST X,' HE DENIES CHARGE": "Dr. J. W. Weinberg of University of Minnesota Is
Accused by House Committee Named 'Scientist X' by House Unit," *The New York Times*, 01
October 1949.

amendment to remain silent in order to not incriminate himself or others.[16] The committee was looking for a certain "scientist X," putatively the person who had handed the atomic secrets to the Soviets. The targets of the committee included Bohm, Lomanitz, Weinberg, Peters, and Frank Oppenheimer, who was brother of J. R. Oppenheimer.

To cut a long story short, we present a summary of the events concerning Bohm's being subpoenaed before the US Congress and the ensuing press coverage. Bohm's case began when he appeared at the HUAC on May 25, 1949, and used his right to remain silent. On August 11, 1950, the House of Representatives cited 54 people, including Bohm, for contempt of Congress for not answering questions from the HUAC committee.[17] On December 4, 1950, a Grand Jury indicted Bohm, together with Lomanitz, Fox, and Steve Nelson, for contempt of Congress. Bohm was immediately arrested and on the same day Princeton University suspended him from "all academic duties" and prevented him from attending classes and using the university libraries. Princeton's decision was accompanied by a declaration by its President, Harold W. Dodds, which anticipated what would happen next, namely the termination of his contract: "His present appointment is for a period ending in June, 1951." Bohm was released on bail and on May 31, 1951, was acquitted from the charges by a federal judge. In June 1951, Princeton, however, did not renew his contract. Like Kafka's protagonist in *The Trial*, Bohm never knew exactly what he was accused of (Fig. 3.4).[18]

How should we analyze the background to these events and Bohm's conduct? First, in order to understand their speed, we must understand that it was the peak of Cold War. In 1947 the Truman Doctrine was announced ending the former military alliance between the US and the Soviet Union; in the mid-1948 the Soviet Union blocked West Berlin; the Soviets detonated their first atomic bomb on 29 August 1949; the Chinese Communists proclaimed the People's Republic of China on 21 September 1949; and the Korean War broke out on 25 June 1950. According to historian Odd A. Westad, "Truman's second-term foreign policy was marked by increasing tension with Soviets, the collapse of a US-supported government in China, and the outbreak of the Korean War. This was the time when the Cold War was militarized, both from a Soviet and American perspective." However, the Cold War should not only be

---

[16]Constitution of the United States, Amendment V, 1789, rev 1992: "No person shall be [...]; nor shall be compelled in any criminal case to be a witness against himself, nor be deprived of life, liberty, or property, without due process of law [...]."

[17]We have a direct evidence of Bohm's reaction to the citation: "Even though I anticipated this trouble, it came as a shock. I am trying to hold up my courage, but I cannot help feeling that the future looks black, because of the madness which is spreading over the country." Bohm to Hanna Loewy [1950], in Talbot (2017, 113).

[18]See *The New York Times* on 26 May 1949, 12 August 1950, 5 December 1950, and 1 June 1951. The historical records of Bohm's case are now well settled. Persecution of Bohm and his colleagues at Berkeley (Bernard Peters, Joseph Weinberg, and Giovanni Rossi Lomanitz) are studied in Shawn Mullet (2008); Princeton's attitudes are analyzed in Russell Olwell (1999); the anti-communist anxieties in American academia in Ellen Schrecker (1986), Jessica Wang (1999), and David Kaiser (2005). Bohm's imprisonment and bail is also recorded in Kojevnikov (2002, 181).

Fig. 3.4 David Bohm reading the newspaper after refusing to testify as to whether or not he was a member of the Communist Party before the House Un-American Activities Committee, 1949. *Credits* Library of Congress, New York World-Telegram and Sun Collection, courtesy AIP Emilio Segrè Visual Archives

seen as tensions on the international scene, it also had far-reaching implications in domestic affairs, both in the US and the U.S.S.R. as well as in many other countries. Still following Westad, "There was never much doubt in the president's mind that the struggle was *both* against the Soviet Union and Communism globally." If in the US the Cold War produced a wider audience for the House Un-American Activities Committee (HUAC) and the phenomenon of McCarthyism, "the alarm that the Cold War created in the United States paled in comparison to the spasms that the Soviet Union and eastern Europe went through. Up to Stalin's death in 1953, denunciations, purges, and show trials were the order of the day."[19]

In the US anxiety related to what was perceived as a Soviet threat triggered a wave of political repression against anybody perceived as a Communist or sympathizers of Communism. This had a huge impact on political and cultural life in the US given the influence of the US Communist Party. Blacklisted Hollywood actors, directors and screenwriters, Hoover's anti-communist obsession, and Senator Joseph McCarthy

---

[19]On the events, see Gaddis (2005). On the analysis of influence of Cold War in domestic affairs, in the US and in the Soviet Union, see Westad (2017, 103–121).

career are a few of the most visible faces of this process.[20] Equally effective was the anxiety towards the people who could have done atomic espionage and thus helped the Soviets to develop their atomic bomb. As remarked by Sam Schweber, this crusade cannot be considered "paranoid" since "a real threat existed. With the opening of Soviet-era state archives, explicit proof of espionage by the USSR became available: the code names and salary receipts of the agents and copies of the sensitive materials of the Manhattan Project and on American diplomatic strategy that were given to the Soviets can be found there." It should be added that the US authorities knew that there had been leakage of information from the Manhattan project to the Soviet Union. This was due to the VENONA project, which surveyed and decrypted messages from the Soviet Embassy in the US to Moscow.[21]

However, still according to Schweber, "the danger was exaggerated and the responses to it […] were out of proportion to the reality of the situation. The harm that was inflicted on the United States by Soviet-sponsored spies did not justify McCarthy's witch hunts and wave of terror. Not only did he wreck individual lives, but he silenced alternative political discourse, thus undermining the foundations of American democracy." Furthermore, Schweber placed all these events not only under the umbrella of the Cold War but also as "a continuation of the struggle that had been waged between conservatives and liberals during the 1930s. The election of Franklin D. Roosevelt had been an important victory for the liberal cause. The conservative forces capitalized on the Soviet threat as a means to regain control in the political arena." Similar lines of reasoning were adopted by Westad, who renewed the historiography on the Cold War presenting it as a grand battle between capitalism and socialism, beginning at the end of the 19th century and going beyond the end of the USSR. Westad also presents a related balance of the effects of McCarthyism on American political life. In his words, "Joseph McCarthy, the demagogic and hyperbolic Wisconsin senator who through his speeches on the Senate floor came to symbolize anti-communist paranoia, did more damage to US interests than any of Stalin's covert operations." Indeed, still according to Westad, "one effect of McCarthyism was that public hysteria made investigations into genuine spy networks more difficult."[22]

Along analogous lines, according to historian of physics David Kaiser, being a theoretical physicist with leftist inclinations in Cold War America was enough to mark anybody as a highly probable target of an anticommunist campaign. For an American layman, the atomic bomb could be reduced to a single equation which could be passed on to the USSR. This supposed that the enemy would immediately

---

[20]Only in recent years has American culture begun to heal from the scars of McCarthyism, with movies such as "Good Night, and Good Luck" and "Trumbo." Even the interest, among historians and philosophers of science, in Bohm's life and works are not fully independent of this new mood.

[21]Schweber (2000, 115–116). On the Soviet-sponsored espionage of the US atomic project, see Rhodes (1995) and Schrecker (2002, 38–42). The updated version of this espionage activity, according the US view, is at the U.S. Department of Energy website dedicated to the history of the Manhattan Project: https://www.osti.gov/opennet/manhattan-project-history/Events/1942-1945/espionage.htm, accessed on 22 August 2018.

[22]Schweber (2000, 115–116), Westad (2017, 120–121).

possess the same weapons America had developed during World War II.[23] We may illustrate Kaiser's point with a few examples concerning the press coverage of David Bohm's case. In 1951, the conservative newspaper *The Mercury* began to publish a series of articles, written by William Bradford Huie, titled "Who Gave Russia the A-BOMB?" In the first of these articles it is stated that "Joe Weinberg delivered a vital formula to Steve Nelson in 1943; the agents *watched* Nelson pass the formula to Peter Ivanov, ..." At that moment, Weinberg had already been identified as "scientist X." However, any serious analysis would consider the concept of one "formula" encapsulating the complex scientific and technological challenge of building the atomic bomb naïve. In the worst case, if true, this leak would signal the existence of an American atomic project to the Soviets. But this had already been suspected by the Soviets and by the Germans and Japanese since American physicists had gone silent, not publishing papers on nuclear fission. In the same vein, Huie continued "it is believed that as early as 1939 the Comintern learned that our experiments in nuclear physics might have military applications." All physicists who were tuned into nuclear physics, the fashionable subject of the 1930s, learned this following the discovery and later explanation of nuclear fission, in December 1938, in Germany, not the US, by Otto Hahn, Fritz Strassmann, Lise Meitner (who by then had fled Nazi Germany), and Otto Robert Frisch.[24] The same article goes on to name the members of the Communist cell at Berkeley, including Bohm, Lomanitz, and Irving David Fox, in addition to Weinberg. Even the regular press stressed the political climate with headlines such as "A-Bomb Scientist Evades Red Query," in *The Washington Post*, or with the infamous picture of him, after leaving the HUAC committee, reading a newspaper with the title "Reds capture all [Shanghai]." Bohm's over exposure to the media is demonstrated by the fact that during these two years his name appeared in print in *The New York Times* at least 13 times.[25]

Bohm's appeal to the Fifth Amendment stirred and still stirs up debate as to whether it was a wise decision or not. Was it the best legal conduct? Did his loyalty to the Communist Party prevail over his loyalty to the US? To answer this question we need to consider both the judiciary technicalities and the context of McCarthyism. Historian Ellen Schrecker explains that at the time Bohm went before HUAC it was no longer safe to appeal for the right of free association as it was this right with regard to the Communist Party which was at stake in the Court. The Fifth Amendment can

---

[23] According to Kaiser (2005, 28), "the early years of the Cold War were not a pleasant time to be an intellectual in the United States, especially if they happened to have a past or present interest in the political left. [...] theoretical physicists emerged as the most consistently named whipping-boys of McCarthyism."

[24] On the plans for voluntary secrecy about nuclear research, particularly led by Leo Szilard and Viktor Weisskopf, see Rhodes (1986, 281–347). On the immediate spread of the news and the implications of the discovery of uranium fission, see Rhodes (1986, 264–275).

[25] William Bradford Huie, "Who Gave Russia the A-BOMB?", *The Mercury*, 412–421, [1951], copy at Bohm Papers, Folder A.120. "A-Bomb Scientist Evades Red Query," *The Washington Post*, 26 May 1949. The picture of David Bohm reading a newspaper; after refusing to testify whether or not he was a member of the Communist Party before the House Un-American Activities Committee is at Library of Congress, New York World-Telegram and Sun Collection, copy also at AIP Emilio Segre Visual Archives. For the over exposure, see the digital archives of *The New York Times*.

be traced back to 17th century England as a "privilege of not having to testify against oneself" to shield dissidents, however, "its use before congressional committees was relatively new and it was unclear just how much protection the Supreme Court would allow it to provide." As a matter of fact, in the context of the anticommunist campaigns, the appeal to this right was a "public relations disaster" as the committees and their supporters claimed that witnesses did not want to cooperate "because they were trying to hide something."[26] Here we arrive at the main point concerning the righteousness of Bohm's conduct. In general McCarthyism expected their targets to name names. This was not specific for suspects of espionage. To name names was part of the game to expose people associating them to Communists. Persecutions would come later and not necessarily in legal terms as indeed economic pressure was more effective than legal procedures. Refusing to name the names of his fellows, including Communists as himself, was the manner Bohm found morally right to act in these circumstances.

David Bohm found support from his fellow physicists throughout this entire process. Graduate students at Princeton, 38 of them, published a letter in *The Washington Post*, titled "Confidence in Bohm," on December 23, 1950. The public letter stated "we wish to record the respect we have felt for Dr. Bohm while he has been at Princeton," and endorsed an official declaration of the university, issued on May 27, 1949, when he had appeared before the HUAC: "he has been regarded by his Princeton colleagues as a thorough American and at no time has there been any reason for questioning his loyalty toward the Government and democratic institutions of the United States." When in prison, money for bail was lent to him by Sam Schweber, then a graduate student at Princeton, who drove him home. In Washington, DC, to appear before the HUAC committee, he was hosted by Edward Condon, who suggested the services of the attorney Clifford Durr. Brazilian graduate students at Princeton, particularly Jayme Tiomno, invited him to go to São Paulo after the university did not renew his contract. Oppenheimer and Albert Einstein wrote letters to the Brazilian authorities supporting him and Oppenheimer advised him to get a passport as soon as possible. Most of his colleagues from the Physics Department at Princeton supported the renewal of his contract by the university, which did not materialize.[27]

Initially the Princeton administration had an ambiguous attitude towards David Bohm. Just after his appearance before the HUAC committee, the university issued a declaration confirming their trust in Bohm's loyalty. After he was indicted for contempt of Congress, because he appealed to the Fifth Amendment to remain silent, the university suspended his contract, with payment, and prohibited him from going on campus. Eventually, the administration did not renew his contract in spite of the approval of the Physics Department. Bohm's recollections were that he felt under pressure from the President Harold W. Dodds, to speak at HUAC naming names, with which he did not comply. In fact, Dodds had in mind not to renew Bohm's contract well before he was cited and later indicted, as Bohm told Hanna Loewy, his

[26]Schrecker (2002).
[27]*The Washington Post*, 23 December 1950. Olwell (1999), Freire Junior (2015).

girlfriend at the time, on March or April 1950: "Oppenheimer happened to talk to Dodds, who indicated that there would be a lot of trouble about my reappointment, and that there was no chance for it at all unless the faculty backed me very strongly." Not all American universities acted in the same manner. The University of Minnesota, for instance, kept Bernard Peters, even when later, due to difficulty getting research funds, he left the US for India and later Denmark. Even nowadays, the wounds caused by Bohm's case and Princeton's conduct towards him, are far from being healed. John Archibald Wheeler, the physicist who took him to Princeton, stated in his autobiography: "the university was gauche in its manner of dealing with Bohm, yet I could sympathize with its goal, to preserve its reputation as a center of unbiased scholarly inquiry, not the home of blind loyalty to one ideology or another. Bohm chose to proclaim neither his independence of thought nor his commitment to Soviet doctrine. Admirably stubborn, he proclaimed only his abhorrence of the Communist witch hunt that was under way."[28]

With the benefit of seventy-year hindsight, what can we conclude from Bohm's case? The ongoing investigation on the information leakages from the Manhattan Project, including information stored in the Soviet archives and the revealing of the VENONA transcripts (The American project to gather and decrypt messages from Soviet agents based in the US), so far, have found no evidence of Bohm's engagement in espionage. More recently Adam Becker, through a FOIA request, applied and obtained access to Bohm's FBI File. No novelty about the suspicion of leakage of atomic secrets through Bohm came out.[29] The Berkeley Radiation laboratory had its activities scrutinized for years. In recent years, documents unearthed in British archives revealed that following information from the FBI the British MI5 surveilled Maurice Wilkins, Bohm's colleague at Berkeley, till 1953 when they dropped the case due to lack of evidence.[30] Espionage, which indeed existed, happened, as far as we know, elsewhere.

Bohm suffered lasting personal and professional damage. The admirable stubbornness, in Wheeler's view, can be seen from another perspective. Others at the apex of McCarthyism conceded in the vain hope of negotiating better treatment. Bohm was stubborn in his principles and courageous enough to endure the consequences of standing by them. Finally, in hindsight, Princeton also lost out. When

---

[28]"PRINCETON PRAISES BOHM UNDER INQUIRY," *The New York Times*, 27 May 1949. On Princeton's decision to not renew, see Olwell (1999). On the meeting with Dodds, Bohm, interviewed by Martin Sherwin, recalled "Dodds called me in. Said I should testify. [First time Bohm had met him]. We talked it over and he sort of kept on intimating that it would be best to testify. 'Well, I said, I couldn't. You know. I couldn't mention names.'" D. Bohm interviewed by M. Sherwin, 15 June 1979, Bohm Papers. On Dodds' early mind, see letter from Bohm to Hanna Loewy [March/April 1950], in Talbot (2017, 99). On Bernard Peters' case, see Schweber (2000, 115–130), Wheeler and Ford (1998, 216).

[29]David Bohm's FBI File, FOIA request by Adam Becker. I am grateful to Becker for sharing this document with me.

[30]"Nobel-winning British scientist accused of spying by MI5, papers reveal," *The Guardian*, 26 August 2010.

he left he was on the way to building a new interpretation for quantum mechanics, which has proven to be valuable in the flourishing research on the foundations of quantum theory, particularly its influence on the appearance of Bell's theorem in the mid-1960s. After Bohm left, Princeton would host another physicist deeply interested in these subjects, Hugh Everett, who would leave physics for a different reason. Princeton already had John von Neumann, Einstein, Eugene Wigner, and Wheeler himself, all interested in the debate on the meaning of quantum physics. Princeton in the late 1950s, with Bohm, Everett, Wheeler, Wigner, and later Abner Shimony, could have become a world center in the research on the foundations of quantum mechanics. It is a counterfactual reasoning as Bohm was expelled from Princeton and from the US.[31]

After Princeton refused to renew his contract, Bohm realized he was highly unlikely to get another job in American academia so he began to look for opportunities abroad, and he was right. Independent of the fairness of the accusations against him, in the context of the Cold War, a physicist with any former Communist affiliation, having worked so closely to the atomic project and remaining loyal to the Communist Party, would not get a job as a physicist. In exile, however, Bohm would not be free from constraints because as a "knowledgeable body," to use Mario Daniels' terminology, he was subject to the US regime of controls on passports and visas. According to Daniels, since the 1940s the US implemented "a differentiated and complex system" embedding "information control in the control over people, restricting circulation to those nationals who have security clearance and curtailing its communication to foreign individuals and entities within and without the US borders."[32] Bohm would experience such controls until the twilight of the Cold War.

Bohm was supported by Oppenheimer and Einstein, and left the US exiling himself for the rest of his life. His first stop in the 1950s was Brazil, then from Brazil he moved to Israel, and from Israel to the UK, where he finally settled. It was after unsuccessfully attempting to work in the UK, in Manchester, that Bohm considered the possibility of exile in Brazil. A small group of Brazilian physicists had graduated from Princeton, among them Jayme Tiomno, a student under John Wheeler and Eugene Wigner in 1950; José Leite Lopes, who had studied under Wolfgang Pauli and Josef Jauch in 1946 and was named a Guggenheim Fellow in 1949; and Walter Schutzer who had completed a Master's degree in 1949. Bohm was one of the readers of Tiomno's doctoral dissertation and served as the chairman of his dissertation committee when Wigner was away. Tiomno invited Bohm to the University of São Paulo. The appointment had the recommendation of Einstein and Oppenheimer and the support at the University of São Paulo of Abrahão de Moraes, then the head of the Physics Department, and Aroldo de Azevedo, an influential geographer. Later, to keep Bohm in his Brazilian position, de Moraes asked Einstein to send letters in favor

---

[31] On a different conduct towards McCarthyism, see Schweber (2000) on the case of Oppenheimer. On Everett, see Freire Junior (2015, 75–139) and Byrne (2010).
[32] Daniels (2019, 35–36).

of an eventual promotion addressed to the highest administrative levels, including President Getúlio Vargas. Bohm arrived in Brazil on October 10, 1951 and left for Israel in early 1955.[33]

3.4   A New Interpretation for the Quantum Theory

Despite the turmoil resulting from the HUAC hearings, Bohm did not give up his research, rather, he diversified the spectrum of his interests. He kept working on plasma and metals as well as on quantum mechanics. However, in the latter it was no longer a matter of working out its best pedagogical presentation. Instead he formulated an entirely new interpretation for this theory to a certain extent in opposition to what he had presented in the textbook *Quantum Theory*. Bohm initially called his work an interpretation of the "quantum theory in terms of 'hidden' variables," and later dubbed it the causal interpretation. The papers presenting this new interpretation were submitted to *Physical review* on July 5, 1951 and published on January 15, 1952. The full work was completed between the late 1950 and June 1951 while Bohm was suspended from Princeton, thus without academic duties. It was improved still in its early stages due to an exchange of letters with Wolfgang Pauli. We shall come back to the exchange as well as the influences and interactions in this early stage as they may shed light on the controversy over the interpretation of the quantum Bohm was working on. We first present Bohm's proposal as it was finally published, then we move to discuss the context of its inception.[34]

3.4.1   The 'Hidden' Variables Interpretation of Quantum Theory[35]

The lengthy paper was entitled "A suggested interpretation of the quantum theory in terms of 'hidden' variables" and was organized in two parts. The technical title hid its far-reaching philosophical implications. Soon both David Bohm and his critics were using "causal interpretation" to label his approach to quantum theory, clarifying Bohm's ambition to restore a kind of determinism analogous to that of classical mechanics. Unlike the early critics of quantum mechanics, Bohm did not just express concern with the usual interpretation of quantum mechanics in terms of a principle

---

[33] Albert Einstein to Patrick Blackett, 17 April 1951, Albert Einstein Archives. Jayme Tiomno, interviewed by the author, 4 August 2003. Record number 816/51 [microfilm], Archives of the Faculdade de Filosofia, Ciências e Letras, Universidade de São Paulo. Abrahão de Moraes did not need to use the letter to President Vargas, it is published in *Estudos avançados*, [São Paulo] 21 (1994).

[34] Bohm (1952a).

[35] The following two subsections are slightly modified versions of Chapter 2 of my book *The Quantum Dissidents* (Freire Junior 2015).

of complementarity. In fact, he built a model for his approach which assumed that an object like an electron is a particle with a well defined path, which means it is simultaneously well defined in both position and momentum, while they cannot be simultaneously measured with arbitrary precision. It is noteworthy that in quantum theory, it is precisely the impossibility of such a simultaneous determination and even definition which breaks with classical determinism, while in classical mechanics the possibility of that simultaneous definition assures the classical deterministic description. Bohm's work had philosophical implications as a consequence of its physical assumptions. According to him, this interpretation "provides a broader conceptual framework than the usual interpretation, because it makes possible a precise and continuous description of all processes, even at the atomic level." More explicitly, he stated that,[36]

> This alternative interpretation permits us to conceive of each individual system as being in a precisely definable state, whose changes with time are determined by definite laws, analogous to (but not identical with) the classical equations of motion. Quantum-mechanical probabilities are regarded (like their counterparts in classical statistical mechanics) as only a practical necessity and not as a manifestation of an inherent lack of complete determination in the properties of matter at the quantum level.

Fully aware of the philosophical implications of his proposal, Bohm concluded the paper by criticizing the usual interpretation of quantum mechanics on philosophical grounds. He accused "the development of the usual interpretation" of quantum theory of being "guided to a considerable extent by the principle of not postulating the possible existence of entities which cannot now be observed," and remarked that "the history of scientific research is full of examples in which it was very fruitful indeed to assume that certain objects or elements might be real, long before any procedures were known which would permit them to be observed directly," the case of the atomistic hypothesis being the best historical example. Bohm also noted that the principle he was criticizing derived from the "general philosophical point of view known during the nineteenth century as 'positivism' or 'empiricism.'" Then he explained to his readers that "a leading nineteenth-century exponent of the positivist view was Mach." Bohm conceded that "modern positivists appear to have retreated from this [Mach's] extreme position." But, he stated, that this position was still reflected "in the philosophical point of view adopted by a large number of modern theoretical physicists." Apart from this philosophical digression, which reflects Bohm's commitment to Marxist philosophy, the philosophical implications of Bohm's proposal concerned not only the recovery of determinism as a mode of description of physical phenomena, but also the adoption of a realist point of view toward physical theories, both discarded by the complementarity view.[37]

---

[36] Bohm (1952a). For the use of the term causal interpretation, see Bohm (1952b), Bohm (1953), and Bohm and Vigier (1954), Bohm (1952a, 166).

[37] Bohm (1952a, pp. 188–189). Bohm's reference to Ernst Mach, criticizing the positivist view, is a shibboleth of his Marxist background as this reference gained currency among Marxists in the first half of the 20th century following the diffusion of *Materialism and Empirio-criticism* (Lenin 1947).

Later in his career, Bohm emphasized that recovering determinism was not his main motivation and that his major dissatisfaction was that "the theory could not go beyond the phenomena or appearances." Building an ontology to explain phenomena would become a permanent goal in Bohm's research with determinism further down in his agenda. However, in the 1950s Bohm did indeed promote the recovery of determinism. In 1951, before the term "causal interpretation" had gained currency in the debates on Bohm's proposal, he himself emphasized it in his first letter to the French astrophysicist and Marxist Évry Schatzman, while looking for allies, such as Jean-Pierre Vigier and Louis de Broglie, to get support for his proposal: "My position in these physical questions is that the world along with all observers who are part of it is objectively real and in principle precisely definable (with arbitrarily high accuracy), and subject to precise causal laws that apply in each individual case and not only statistically."[38] The debate triggered by this, with the coinage of the term causal interpretation, brought the recovery of determinism to the foreground of the quantum debates.

To illustrate the strength of the attachment of Bohm and his collaborators to the philosophical priority of causality, we can make reference to the work that he and Jean-Pierre Vigier did in 1954, changing Bohm's original model slightly. In this work, they embedded the electron in a fluid undergoing "very irregular and effectively random fluctuation" in its motion. While these fluctuations could be explained by either a deterministic or a stochastic description, Bohm and Vigier framed them into the causal interpretation approach, giving their paper the title "Model of the causal interpretation of quantum theory in terms of a fluid with irregular fluctuations."[39]

Bohm not only suggested a new conceptual and philosophical framework, he also raised the stakes by suggesting his approach could be fruitful in new domains of physics. He suggested that "modifications can quite easily be formulated in such a way that their effects are insignificant in the atomic domain [...] but of crucial importance in the domain of dimensions of the order of 10^{-13} cm." Bohm was indeed referring to intra-nuclear distances, an area in which there was a proliferation of discoveries of new particles requiring the development of new methods in quantum field theories. Bohm's promises, however, were as appealing as vague, saying that "it is thus entirely possible that some of the modifications describable in terms of our suggested alternative interpretation, but not in terms of the usual interpretation, may be needed for a more thorough understanding of phenomena associated with very small distances." The promise of fulfilling such an expectation was then postponed: "we shall not, however, actually develop such modifications in any detail in these papers."[40]

---

[38]Bohm (1987, 33). David Bohm to Évry Schatzman [1951]. The letter is transcribed in Besson (2018, 307–308). The original is deposited in the Archives de la Bibliothèque de l'Observatoire, Paris.

[39]Bohm and Vigier (1954).

[40]Bohm (1952a, 166).

In Bohm's original model, electrons undergo physical influences both from potentials, such as electromagnetic potentials, and from a new potential resulting from mathematical manipulations of the Schrödinger equation, which Bohm labeled the "quantum potential." Technically, this new potential arises when one exploits analogies between the Schrödinger equation of quantum theory and the Hamilton-Jacobi equation of classical mechanics. To illustrate, let us take an electron, as Bohm did, with well defined positions described by a function of the form $\psi = R \exp(iS/\hbar)$, which must satisfy the Schrödinger equation, and let us call $R(\mathbf{x})^2 = P(\mathbf{x})$. After some mathematical manipulations we get Eqs. (3.1) and (3.2) resulting from Bohm's approach.

$$\partial P/\partial t + \nabla(P\nabla S/m) = 0 \qquad (3.1)$$

$$\partial S/\partial t + (\nabla S)^2/2m + V + U = 0 \qquad (3.2)$$

where

$$U = -\left(\hbar^2/2m\right)(\nabla^2 R/R)$$

Bohm then further exploited these analogies by suggesting that electrons have a well defined momentum $\mathbf{p} = \nabla S(\mathbf{x})$. The same analogies suggest that the "extra" term U in Eq. (3.2) may be interpreted as the action of a "quantum potential" on electrons, in addition to the potentials known from classical physics, such as electromagnetic potentials. Furthermore, according to this model, Eq. (3.1) is a continuity equation, and Bohm suggests that we take $P = |\psi(x)|^2$, where ψ is the solution of the Schrödinger equation, to assure the conservation of the probability density of an ensemble of particle positions. As remarked by historian of physics Max Jammer, "Bohm interprets $[\psi^*\psi]$ as the probability of the particle's *being* at the position defined by the argument x of $\psi(x)$ and not, as Born conceived it, as the probability of *finding* the particle at that position if performing a suitable measurement." Bohm's model of electrons has well defined positions as well as momenta, thus they have continuous and well defined trajectories. These p's and x's are the hidden variables in Bohm's models. They are "hidden" when compared to standard quantum mechanics as Heisenberg's uncertainty relations do not allow the simultaneous precise definitions of positions and momenta. Indeed, these hidden variables had been introduced by Bohm from the beginning of his reasoning when he assumed an electron as a quantum object with well defined position and then suggested it also had a well defined momentum given by $\mathbf{p} = \nabla S(\mathbf{x})$. Heisenberg's uncertainty relations and Bohr's complementarity were not compatible with such assumptions. Bohr's complementarity view stated that position and momentum were complementary but mutually exclusive physical variables, the proper use of each being constrained by the conditions of measurement. Later on, however, the physicist John S. Bell, a supporter of Bohm's proposal, would consider Bohm to have made an unhappy choice in the term "hidden variables." Bell would remark that complementarity is the interpretation which hides one of the complementary variables as they could not be considered images of the phenomena.[41]

In order to obtain models able to produce the same results as quantum mechanics, Bohm needed to ascribe well defined positions and momenta to the measurement devices too. Thus, from the Hamiltonian (kinetic plus potential energies) of the coupling between such devices and the micro systems, observable results could be predicted. Bohm used these models to carry out detailed calculations of a number of different problems, for instance, stationary states, transitions between stationary states (including scattering problems), the Einstein-Podolsky-Rosen *Gedankenexperiment*, and photoelectric and Compton effects. To achieve results compatible with those from quantum mechanics, Bohm modeled light as electromagnetic waves. In all these problems he found the results predicted by the usual mathematical formalism of quantum theory.[42]

Bohm's achievement was not a minor one. He was able to build an approach to quantum theory leading to the same predictions as usual quantum mechanics and develop the first alternative interpretation to the dominant complementarity view among physicists. Bohm was fully aware of this. "His excitement was great: 'To think that I was the one to have seen this,'" are the recollections of Miriam Yevick from their meeting in early 1951. This empirical equivalence was dependent on adopting hidden variables in the system and in the measurement device, which was an improvement on his initial approach. This was done in reaction to criticisms made by Wolfgang Pauli as we will see later. To be more precise, Bohm's approach was equivalent to non-relativistic quantum mechanics as his electron model, for instance, did not have "spin."[43]

That Bohm's approach was unable to deal with relativistic systems is clear from the equation of the quantum potential. Indeed, if you take a system with two electrons the quantum potential tells us that an interaction could propagate from one electron to the other instantaneously without any time dependency. This would not have been considered a major flaw when Bohm published his papers if one recalls that in the historical process of the creation of quantum physics, non-relativistic equations came first and relativistic generalizations a little later. At any rate, critics would ask Bohm for these generalizations and Bohm would promise that they were under way.

---

[41]Jammer (1988, 693). In Bell's words (Bell 2004, p. 201), "absurdly, such theories are known as 'hidden variable' theories. Absurdly, for there is not in the wavefunction that one finds an image of the visible world, and the results of experiments, but in the complementary 'hidden'(!) variables." I am thankful to Michael Kiessling for pointing out these remarks to me.

[42]Jammer (1988), Bohm (1952a, 183).

[43]Yevick (2012, 141). Miriam Yevick was a mathematician, the fifth woman to obtain a PhD in Mathematics at MIT, who befriended Bohm for a long time, maintaining extensive correspondence while he stayed in Brazil. This has now been edited and published in Talbot (2017). Yevick had been fascinated by Bohm since their first meeting in January 1948 (Yevick 2012, 140–141): "I held out my hand, looked up, and was struck by a *Coup de Foudre*. Our eyes met, our glances penetrated each others' and would not let go. This happened whenever we met from then on. [...] Truly, I became his disciple for the rest of my life." Miriam was married to the physicist George Yevick. The couple remained close to Bohm.

Bohm's papers also raised other philosophical and technical issues. Empirically equivalent to standard quantum mechanics, Bohm's would be a nice example of what philosophers call the underdetermination of theories by empirical data. According to the philosopher Paul Feyerabend, after Bohm's work, "it follows that neither experience nor mathematics can help if a decision is to be made between wave mechanics and an alternative theory which agrees with it in all those points where the latter has been found to be empirically successful." This philosophical thesis, also called the Duhem-Quine thesis, a reference to the scientist and philosopher Pierre Duhem and the philosopher Willard Van Orman Quine, was well set in logical terms but it was, and is, somewhat unpleasant for physicists to accept that some of their best theories are not the only possible description of phenomena.[44] Finally, Bohm's approach was a practical example showing that something was wrong with von Neumann's mathematical proof against the possibility of introducing hidden variables in quantum mechanics. This proof had been in von Neumann's presentation of the mathematical foundations of quantum mechanics in 1932. Bohm was fully aware of this in his approach, making it explicit in his papers, and he would attentively follow von Neumann's reactions to his proposals. Finally, as we have already noted, Bohm did not ignore the philosophical debate implied by his proposals as he not only defended his approach with both technical and conceptual arguments, but also accused supporters of the standard interpretation of being the 20th-century equivalents to the anti-atomists of the 19th century.[45]

3.4.2 Background to the Inception of the 'Hidden' Variables Interpretation

Bohm would later acknowledge at least two influences on the origins of his proposal of a new interpretation of quantum mechanics: a discussion with Albert Einstein at the Institute of Advanced Studies in Princeton after the book *Quantum Theory* was published and the reading of a paper by a Soviet physicist criticizing the complementarity view for its idealistic and subjectivist inclinations. These influences resonated with the views Bohm had been nurturing the previous decade, a realistic worldview, as expressed in the book *Quantum Theory*, and his engagement with a Marxist theoretical framework in thinking about society and nature. As told by historian Max Jammer,[46]

> Stimulated by his discussion with Einstein and influenced by an essay which, as he told the present author, was "written in English" and "probably by Blokhintsev or some other Russian theorist like Terletzkii," and which criticized Bohr's approach, Bohm began to study the possibility of introducing hidden variables.

---

[44]Feyerabend (1960, 325). On the Duhem-Quine thesis, see Stanford (2017) and Harding (1976). On quantum mechanics as an illustration of this thesis, see Cushing (1994).

[45]von Neumann (1955), Bohm (1952a, pp. 187–188).

[46]Jammer (1974, 279).

Later Jammer reiterated this story in a kind of Festschrift for Bohm's 70th birthday. Bohm never contested it. This information, however, raised a doubt, as Jammer himself had noted. "Bohm [had] forgotten the exact title and author of this paper" and there was no paper either by Blokhintsev or by Terletzkii published in English before Bohm's shift to the causal interpretation. Indeed, the papers by the Soviets criticizing complementarity were first translated and published in French, not in English, and this in 1952, while Bohm's shift to the causal interpretation occurred in 1951.[47]

The confusion may be explained if one takes into account that criticism towards complementarity and what was seen as subjectivist tendencies were part of the ideological battles in the Soviet Union. The historian of Soviet science Loren Graham considered the late 1940s the beginning of what he called "the age of the banishment of complementarity" in the USSR, as part of the Zhdanovshchina, "the most intense ideological campaign in the history of Soviet scholarship." The Soviet criticisms quickly reached Marxist physicists throughout the world. It was Léon Rosenfeld, a Belgian physicist working in Manchester and close to Niels Bohr, who warned him of these developments as early as 1949. Rosenfeld was both a Marxist and a supporter of Bohrian complementarity and would become a bulldog in the following years defending complementarity from attacks from the Marxist milieu. On May 31, 1949, he wrote to Bohr: "[I am] just writing an article on 'Komplementaritet og modern Rationalisme' in order to clear up the various misunderstandings which arise when one tries to mix complementarity in all possible sorts of mysticism, whatever it is a question of idealism as with Eddington and others, or about the Russian Pseudo-Marxism. These many 'ismes' are surely too tedious to [you] but [I] feel that one cannot any longer content oneself by ignoring that nonsense."[48] In this context it is not implausible that free translations from the Soviet papers may have circulated among Marxist intellectual circles in the West before their publication. Plausible as this is, unfortunately, we have no documentary evidence to support it. Archival documents unearthed since then have not been able to corroborate Jammer's interesting statement.

Indeed, most of Bohm's personal papers from his time in the US did not survive, nor did he keep copies of his papers or correspondence. For the particular period he was working on the hidden-variables interpretation, late 1950 and the first half of 1951, we only have his letters to Pauli. Letters from him to some friends and fellow physicists have surfaced but they concern the times either before he began to work on the hidden-variables paper or after its submission to *Physical review*, thus they fail to document the personal and intellectual environment at the time of his move to the

---

[47]Jammer (1974, 279, footnote 63; 1988, p. 692), Blokhintsev (1952), Terletsky (1952).

[48]For Rosenfeld's biography, see Jacobsen (2012). On the debates on the quantum theory in the former USSR, see Graham (1987, pp. 325 and 328) and Kojevnikov (2004). Rosenfeld to Bohr, 31 May 1949, Bohr Scient. Corr, Archive for the History of Quantum Physics, Niels Bohr Archive, Copenhagen.

new interpretation of quantum mechanics.[49] From one of these few letters we can see the time when Bohm had not yet thought of suggesting an interpretation alternative to that of complementarity. In mid-June 1950, Bohm wrote to Hanna Loewy: "My plans are to apply for a Guggenheim to go to either England or Denmark in Summer 1951. I can work with Massey in England, or with Niels Bohr in Denmark. The latter work sounds more interesting as it would involve writing a book on the philosophy of the quantum theory. As for my present book, that has finally been edited for the last time and sent to the printer. Galleys should start coming in July."[50] As an exercise in counter-factual irony one may wonder what would have happened to the history of the interpretations of quantum mechanics if Bohm's expectations had materialized and he had gone to Copenhagen to write a book on the philosophy of quantum mechanics along the lines of Niels Bohr's views instead of staying in Princeton and writing a paper against Bohr's views.

The intersection of Bohm's political persecution and his move towards a new interpretation of quantum theory has attracted the attention of other historians. Christian Forstner has suggested that isolation from Princeton and the American community of physicists was influential in Bohm's abandoning the standard interpretation of quantum physics and adopting a heterodox interpretation. Again, unfortunately, historians have to deal with the scant documentary evidence around the circumstances of these events in the crucial months between his appearance at the HUAC in 1949, the publication of his book *Quantum Theory*, and the completion of his causal interpretation in the middle of 1951. As such, mapping the influences and motivations of his move toward the causal interpretation is just conjecture.[51]

Einstein's conversation with Bohm after the publication of *Quantum Theory* was the start of a relationship which would last until Einstein's death in 1955. Ironically, as it may seem, Einstein would support Bohm on several grounds except in defense of Bohm's approach to quantum mechanics. We shall see some of their exchanges in the next chapter.

---

[49]The David Bohm Papers are deposited at Birkbeck College, University of London. From the period a little before and after leaving the U.S., there is a meaningful correspondence with Einstein; Melba Philips, an American physicist and friend of Bohm; Hanna Loewy and Miriam Yevick, his friends. The letters to Philips, Loewy, and Yevick are transcribed and published in a volume with critical apparatus by Talbot (2017). Most of the correspondence with Wolfgang Pauli, relevant for the period prior to his departure from the U.S. and during the elaboration of his paper on the causal interpretation, was recovered and published by Karl von Meyenn in the collection dedicated to Pauli's correspondence (Pauli and Meyenn 1996, 1999). At Rosenfeld Papers, Niels Bohr Archive, Copenhagen, there are plenty of letters concerning the battle Rosenfeld fought in defense of complementarity and against the hidden-variables; most of them are cited in my *The Quantum Dissidents* (Freire Junior 2015). More recently, a batch of letters between Bohm and the French astrophysicist Evry Schatzman was unearthed by Virgile Besson at Schatzman's papers, Observatoire de Paris. These letters are appended to his doctoral dissertation (Besson 2018) where Besson analyses the early connection between Bohm and the French team around Louis de Broglie and Jean-Pierre Vigier.

[50]Bohm to Hanna Loewy, mid-June 1950, is in Talbot (2017, 110).

[51]Forstner (2008).

3.4.3   Bohm, de Broglie, and Pauli—Conceptual Issues and Disputes About Priorities

While writing his paper on the 'hidden' variables Bohm was unaware of previous work by Louis de Broglie along analogous lines. What we are able to reconstruct about how Bohm reacted when informed of de Broglie's works sheds light on the kind of technical problems he had to solve in order to make his proposal consistent. It is also illuminating regarding the disputes and alliances in the controversy over the foundations of quantum physics. Last but not least, as Wolfgang Pauli was one of the people to warn Bohm about de Broglie's works and as their exchange is one of the most relevant for the early debate on the causal interpretation, it is interesting to see their discussion in some detail. Before Bohm's papers appeared in print, Einstein and Pauli informed him that de Broglie had suggested a similar approach at the 1927 Solvay conference, which Bohm had not known about. Pauli had criticized de Broglie's approach when first proposed and de Broglie had reacted by giving up his idea. Now Bohm had to face the same objections. Pauli had argued that de Broglie's proposal fitted Max Born's probabilistic interpretation of the ψ function only for elastic collisions. In the case of inelastic scattering of particles by a rotator, a problem Enrico Fermi had solved in 1926, de Broglie's idea was incompatible with assigning stationary states to a rotator, before and after the scattering. Pauli had considered this failure intrinsic to de Broglie's picture of particles with definite trajectories in space-time, an approach de Broglie had called the "pilot wave", which means particles with well defined paths ruled by waves coming from the Schrödinger equation.[52]

Pauli addressed his criticisms considering a draft version of Bohm's paper, which Bohm subsequently corrected. This draft has not survived, but an indication of the corrections has. In response to Pauli's criticisms Bohm wrote: "I hope that this new copy will answer some of the objections to my previous manuscript ... to sum up my answer to your criticisms ... I believe that they were based on the excessively abstract assumptions of a plane wave of infinite extent for the electrons' Ψ function. As I point out in Sect. 7 of paper I, if you had chosen an incident wave packet instead, then after the collision is over, the electron ends up in one of the outgoing wave packets, so that a stationary state is once more obtained." Initially Pauli did not read the second manuscript as he considered it too long, which angered Bohm. He rebuked Pauli: "If I write a paper so 'short' that you will read it, then I cannot answer all of your objections. If I answer all of your objections, then the paper will be too 'long' for you to read. I really think that it is your duty to read these papers

---

[52]Einstein's remark is in Paty (1993). Bohm to Pauli, [July 1951], in Pauli and Meyenn (1996, pp. 343–345). Most of Pauli's letters to Bohm did not survive; we infer their contents from Bohm's replies. Bohm to Karl von Meyenn, 2 December 1983, ibid., on 345. Broglie's pilot wave and Pauli's criticisms are in Institut International de Physique Solvay (1928, pp. 105–141 and 280–282). See also Bacciagaluppi and Valentini (2009), which is a critical edition (with a careful analysis as introduction) of the proceedings of the Solvay conference, with an English translation of the Conference proceedings, the originals of which were published in French.

carefully." As a precaution, he summarized his views and the improvements in the own letters, in addition to send enclosed the new draft[53]:

> In the second version of the paper, these objections are all answered in detail. The second version differs considerably from the first version. In particular, in the second version, I do not need to use "molecular chaos." You refer to this interpretation as de Broglie's. It is true that he suggested it first, but he gave it up because he came to the erroneous conclusion that it does not work. The essential new point that I have added is to show in detail (especially by working out the theory of measurement in paper II) that his interpretation leads to all of the results of the usual interpretation. Section 7 of paper I is also new [transitions between stationary states—the Franck-Hertz experiment], and gives a similar treatment to the more restricted problem of the interaction of two particles, showing that after the interaction is over, the hydrogen atom is left in a definite "quantum state" while the outgoing scattered particle has a corresponding definite value for its energy.

Eventually, Pauli analyzed Bohm's papers as well as the letters. He conceded that Bohm's model was logically consistent, which was recognition of Bohm's work: "I do not see any longer the possibility of any logical contradiction as long as your results agree completely with those of the usual wave mechanics and as long as no means is given to measure the values of your hidden parameters both in the measuring apparatus and in the observed system." Pauli ended with a challenge, related to Bohm's promise of applying his approach to new domains such as high energy physics: "as far as the whole matter stands now, your 'extra wave-mechanical predictions' are still a check, which cannot be cashed." Pauli never ceased to oppose the hidden variable interpretation and would formulate new objections, as we will see later. For Bohm, however, Pauli's challenge now was less pressing than de Broglie's.[54]

Before 1927, Louis de Broglie had the idea of a "double solution," in which the waves of Schrödinger equation pilot the particles, which are singularities of the waves. Just before the meeting of the Solvay council on October 24–29, 1927 he gave up this idea because of its mathematical difficulties and presented his report to the meeting with just the "pilot wave" proposal. The particles were reduced to objects external to the theory. After the 1927 meeting he adhered to the complementarity interpretation. Bohm was right in remarking that de Broglie had not carried his ideas to their logical conclusion, but de Broglie surely had a share in the idea of hidden variables in quantum mechanics. Bohm resisted this. He suggested the following interesting analogy which expresses the silent dispute about priorities between the two physicists, the young American and the elder Frenchman: "If one man finds a diamond and then throws it away because he falsely concludes that it is a valueless stone, and if this stone is later found by another man who recognize its true value, would you not say that the stone belongs to the second man? I think the same applies to this interpretation of the quantum theory."[55] Through the correspondence where Bohm approached Schatzman, in order to interact with Vigier, who was a

---

[53] Bohm to Pauli, July 1951, Summer 1951, October 1951, 20 November 1951 (Pauli and Meyenn 1996, pp. 343–346, 389–394, and 429–462).

[54] Pauli to Bohm, 3 December 1951, plus an appendix (Pauli and Meyenn 1996, pp. 436–441).

[55] For the evolution of de Broglie's ideas, see Broglie (1956, pp. 115–143). Bohm to Pauli, October 1951, *op. cit.*

de Broglie assistant, we know that de Broglie had complained to Bohm about the lack of acknowledgment of his priority: "I recently received a letter from de Broglie, in which he took great pains to claim credit for the ideas." Bohm's answer was a statement about what in his own work surpassed de Broglie's early work: "My answer to him was to admit that he suggested the method in 1926, but to point out that because he did not carry it to its logical conclusion he came to the erroneous conclusion that the idea does not work." Furthermore, "the main new element added by me is to show in detail (especially constructing a theory of measurements) that this interpretation leads to all of the experimental predictions given by the usual interpretation." To Schatzman, but not to de Broglie, he also presented the same analogy of the diamond: "With regard to people like de Broglie who had the idea for a similar interpretation of quantum theory, my attitude is that if a man finds a diamond and not realizing its value throws it away, then the stone belongs to the first person who finds it again and does realize its value."[56]

In the end Bohm adopted a diplomatic stance, suggested by Pauli, to recognize de Broglie's contribution while maintaining the superiority of his own work: "I have changed the introduction of my paper so as to give due credit to de Broglie, and have stated that he gave up the theory too soon (as suggested in your letter)." In addition to changing the introduction, he added "a discussion of interpretations of the quantum theory proposed by de Broglie and Rosen" and rebutted Pauli's criticisms. By the time Bohm's papers appeared in print, de Broglie had returned to his old causal approach reviving the idea of "double solution" with his assistant Jean-Pierre Vigier. They would become the most important of Bohm's allies in the hidden-variable campaign.[57]

3.4.4 Leaving Princeton for São Paulo

In early October 1951 Bohm left the United States for Brazil to assume a chair at the Universidade de São Paulo. Bohm's mood was haunted by conflicting aspects. He thought he had finally produced a very original work with the hidden-variables paper, he was aware of the obstacles it would face and was willing and encouraged to fight in its defense. However, he was leaving the US as a kind of political exile due to the anti-communist anxiety which had assaulted his own country. He had no expectations to come back soon. Furthermore, he suffered from loneliness. The previous year he had proposed marriage to Hanna Loewy inviting her to join him on the Brazilian trip, "I certainly hope that it is possible for us to go to Brazil." But by June 1950, she had changed her mind. "It is ironical that just when I want to marry you, you are not interested in "losing your freedom," were Bohm's words to her.[58]

---

[56]Bohm to Schatzman, [1951], in Besson (2018, 307–308). Besson (2018) is also a source for the engagement of de Broglie and Vigier with the 'hidden' variables interpretation.

[57]Bohm to Pauli, 20 November 1951, *op. cit.* (Bohm 1952a, pp. 191–193).

[58]For Bohm's letters to Hanna Loewy, see Talbot (2017, 102 and 110).

References

Bacciagaluppi, G., Valentini, A.: Quantum Theory at the Crossroads: Reconsidering the 1927 Solvay Conference. Cambridge University Press, Cambridge (2009)

Bardeen, J., Pines, D.: Electron-phonon interaction in metals. Phys. Rev. **99**(4), 1140–1150 (1955)

Bell, J.S.: Speakable and Unspeakable in Quantum Mechanics: Collected Papers on Quantum Philosophy. With an Introduction by Alain Aspect. Cambridge University Press, Cambridge (2004)

Besson, V.: L'interprétation causale de la mécanique quantique: biographie d'un programme de recherche minoritaire (1951–1964). Ph.D. dissertation, Université Claude Bernard Lyon 1 and Universidade Federal da Bahia (2018)

Blokhintsev, D.I.: Critique de la conception idéaliste de la théorie quantique. Questions scientifiques—Physique, pp. 95–129. Les éditions de la nouvelle critique, Paris (1952)

Bohm, D.: Quantum Theory. Prentice-Hall, New York (1951)

Bohm, D.: Proof that probability density approaches in causal interpretation of the quantum theory. Phys. Rev. **89**(2), 458–466 (1953)

Bohm, D.: A suggested interpretation of the quantum theory in terms of hidden variables—I & II. Phys. Rev. **85**(2), 166–179–180–193 (1952a)

Bohm, D.: Reply to a criticism of a causal re-interpretation of the quantum theory. Phys. Rev. **87**(2), 389–390 (1952b)

Bohm, D.: Causality and Chance in Modern Physics. Routledge and Paul, London (1957)

Bohm, D.: The Special Theory of Relativity. W.A. Benjamin, New York (1965)

Bohm, D.: Wholeness and the Implicate Order. Routledge & Kegan Paul, London (1980)

Bohm, D.: Hidden variables and the implicate order. In: Hiley, B., Peat, D. (eds.) Quantum Implications: Essays in Honour of David Bohm, pp. 33–45. Routledge, London (1987)

Bohm, D., Gross, E.: Theory of plasma oscillations. A. Origin of medium-like behavior. Phys. Rev. **75**(12), 1851–1864 (1949a)

Bohm, D., Gross, E.: Theory of plasma oscillations. B. Excitation and damping of oscillations. Phys. Rev. **75**(12), 1864–1876 (1949b)

Bohm, D., Pines, D.: A collective description of electron interactions. 1. Magnetic interactions. Phys. Rev. **82**(5), 625–634 (1951)

Bohm, D., Pines, D.: A collective description of electron interactions. 3. Coulomb interactions in a degenerate electron gas. Phys. Rev. **92**(3), 609–625 (1953)

Bohm, D., Hiley, B.J.: The Undivided Universe: An Ontological Interpretation of Quantum Theory. Routledge, London (1993)

Bohm, D., Vigier, J.P.: Model of the causal interpretation of quantum theory in terms of a fluid with irregular fluctuations. Phys. Rev. **96**(1), 208–216 (1954)

Bohr, A., Mottelson, B.R., Pines, D.: Possible analogy between the excitation spectra of nuclei and those of the superconducting metallic state. Phys. Rev. **110**(4), 936–938 (1958)

Broglie, L.: Nouvelles Perspectives en Microphysique. Albin Michel, Paris (1956)

Byrne, P.: The Many Worlds of Hugh Everett III: Multiple Universes, Mutual Assured Destruction, and the Meltdown of a Nuclear Family. Oxford University Press, New York (2010)

Corson, E.M.: Quantum theory. Phys. Today **5**(2), 23 (1952)

Cushing, J.: Quantum Mechanics—Historical Contingency and the Copenhagen Hegemony. The University of Chicago Press, Chicago (1994)

Daniels, M.: Restricting the transnational movement of "knowledgeable bodies"—the interplay of US visa restrictions and export controls in the Cold War. In: Krige, J. (ed.) How Knowledge Moves—Writing the Transnational History of Science and Technology, pp. 35–61. Chicago University Press, Chicago (2019)

Feyerabend, P.: Professor Bohm's philosophy of nature. Br. J. Philos. Sci. **10**(40), 321–338 (1960)

Forstner, C.: The early history of David Bohm's quantum mechanics through the perspective of Ludwik Fleck's thought-collectives. Minerva **46**(2), 215–229 (2008)

Freire Junior, O.: The Quantum Dissidents—Rebuilding the Foundations of Quantum Mechanics 1950–1990. Springer, Berlin (2015)

Gaddis, J.L.: The Cold War: A New History. Penguin, New York (2005)

Graham, L.R.: Science, Philosophy, and Human Behavior in the Soviet Union. Columbia University Press, New York (1987)

Gross, E.P.: Collective variables in elementary quantum mechanics. In: Hiley, B.J., Peat, F.D. (eds.) Quantum Implications: Essays in Honour of David Bohm, pp. 46–65. Routledge & Kegan Paul, London (1987)

Harding, S.G. (ed.): Can Theories be Refuted? Essays on the Duhem-Quine Thesis. Synthese Library, vol. 81. D. Reidel Publishing Co., Dordrecht (1976)

Heisenberg, W.: The development of the interpretation of the quantum theory. In: Pauli, W. (ed.) Niels Bohr and the Development of Physics; Essays Dedicated to Niels Bohr on the Occasion of his Seventieth Birthday, pp. 12–29. McGraw-Hill, New York (1955)

Hoddeson, L., Schubert, H., Heims, S.J., Baym, G.: Collective phenomena. In: Hoddeson, L., Braun, E., Teichmann, J., Weart, S. (eds.) Out of the Crystal Maze—Chapters from the History of Solid-State Physics, pp. 489–616. Oxford University Press, New York (1992)

Howard, D.: Einstein on locality and separability. Stud. Hist. Philos. Sci. 16, 171–201 (1985)

Howard, D.: Einstein's philosophy of science. In: Zalta, E.N. (ed.) The Stanford Encyclopedia of Philosophy, Fall 2017 edn. https://plato.stanford.edu/archives/fall2017/entries/einstein-philscience/ (2017)

Inglis, D.R.: Quantum theory. Am. J. Phys. 20, 522 (1952)

Institut International de Physique Solvay: Electrons et photons—Rapports et discussions du Cinquième Conseil de Physique tenu à Bruxelles du 24 au 29 octobre 1927. Gauthier-Villars, Paris (1928) [English translation in Bacciagaluppi and Valentini 2009]

Jacobsen, A.: Léon Rosenfeld—Physics, Philosophy, and Politics in the Twentieth Century. World Scientific, Singapore (2012)

Jammer, M.: The Philosophy of Quantum Mechanics—The Interpretations of Quantum Mechanics in Historical Perspective. Wiley, New York (1974)

Jammer, M.: David Bohm and his work on the occasion of his 70th-birthday. Found. Phys. 18(7), 691–699 (1988)

Kaiser, D.: The atomic secret in red hands? American suspicions of theoretical physicists during the early Cold War. Representations 90, 28–60 (2005)

Kojevnikov, A.: David Bohm and collective movement. Hist. Stud. Phys. Biol. Sci. 33, 161–192 (2002)

Kojevnikov, A.: Stalin's Great Science: The Times and Adventures of Soviet Physicists. Imperial College Press, London (2004)

Le Couteur, K.J.: Principles of quantum theory. Nature 171, 276 (1953)

Lenin, V.I.: Materialism and Empirio-Criticism; Critical Comments on a Reactionary Philosophy. Foreign Languages Publishing House, Moscow (1947)

Mehra, J.: The Beat of a Different Drum: The Life and Science of Richard Feynman. Clarendon, Oxford (1994)

Mullet, S.K.: Little man: four junior physicists and the red scare experience. PhD dissertation, Harvard University (2008)

Nozières, P., Pines, D.: The Theory of Quantum Liquids. Perseus, Cambridge, MA (1990)

Olwell, R.: Physical isolation and marginalization in physics—David Bohm's Cold War exile. ISIS 90(738–756) (1999)

Paty, M.: Sur les 'variables cachées' de la mécanique quantique—Albert Einstein, David Bohm et Louis de Broglie. La pensée 292, 93–116 (1993)

Pauli, W., Meyenn, K.V.: Wissenschaftlicher Briefwechsel mit Bohr, Einstein, Heisenberg u. a. Band IV Teil I 1950–1952. Springer, Berlin (1996)

Pauli, W., Meyenn, K.V.: Wissenschaftlicher Briefwechsel mit Bohr, Einstein, Heisenberg u. a. Band IV Teil II 1953–1954. Springer, Berlin (1999)

Pines, D.: A collective description of electron interactions: IV. Electron interaction in metals. Phys. Rev. **92**, 625–636 (1953)

Pines, D.: The collective description of particle interactions: from plasmas to the helium liquids. In: Hiley, B.J., Peat, F.D. (eds.) Quantum Implications: Essays in Honour of David Bohm, pp. 66–84. Routledge & Kegan Paul, London (1987)

Pines, D., Bohm, D.: A collective description of electron interactions. 2. Collective vs individual particle aspects of the interactions. Phys. Rev. **85**(2), 338–353 (1952)

Rhodes, R.: The Making of the Atomic Bomb. Simon & Schuster, New York (1986)

Rhodes, R.: Dark Sun—The Making of the Hydrogen Bomb. Simon & Schuster, New York (1995)

Schrecker, E.: No Ivory Tower: McCarthyism and the Universities. Oxford University Press, New York (1986)

Schrecker, E.: The Age of McCarthyism—A Brief History with Documents. Bedford/St. Martins's, Boston (2002)

Schweber, S.S.: QED and the Men Who Made It: Dyson, Feynman, Schwinger, and Tomonaga. Princeton University Press, Princeton, NJ (1994)

Schweber, S.S.: In the Shadow of the Bomb: Bethe, Oppenheimer, and the Moral Responsibility of the Scientist. Princeton University Press, Princeton, NJ (2000)

Stanford, K.: Underdetermination of scientific theory. In: Zalta, E.N. (ed.) The Stanford Encyclopedia of Philosophy, Winter 2017 edn. https://plato.stanford.edu/archives/win2017/entries/scientific-underdetermination/ (2017)

Talbot, C. (ed.): David Bohm: Causality and Chance, Letters to Three Women. Springer, Berlin (2017)

Terletsky, I.P.: Problèmes du développement de la théorie quantique. Questions scientifiques—Physique, pp. 131–146. Les éditions de la nouvelle critique, Paris (1952)

Von Neumann, J.: Mathematische Grundlagen der Quantenmechanik. Julius Springer, Berlin (1932)

Von Neumann, J.: Mathematical Foundations of Quantum Mechanics. Princeton University Press, Princeton, NJ (1955)

Wang, J.: American Science in an Age of Anxiety: Scientists, Anticommunism, and the Cold War. University of North Carolina Press, Chapel Hill, NC (1999)

Westad, O.A.: The Cold War—A World History. Basic Books, New York (2017)

Wheeler, J.A., Ford, K.: Geons, Black Holes, and Quantum Foam. Norton, New York (1998)

Yevick, M.L.: A Testament for Ariela. Blue Tread Communications, Accord, NY (2012)

Chapter 4
The Long Campaign for the Causal Interpretation (1952–1960). Brazil, Israel, and the U.K.

When David Bohm left the US for Brazil in early October 1951, he would never have imagined that his recently finished paper on the hidden-variable interpretation of quantum mechanics would lead him to a decade-long campaign for the then so-called causal interpretation of this physical theory. These activities would absorb almost all of his energies in physics and a more philosophically inclined scientist would emerge, particularly with the publication of his book *Causality and Chance* in 1957. Only in the late 1950s and early 1960s would his engagement in defense of this interpretation fade. However, his quest for understanding quantum theory would survive in different ways. The ghosts of McCarthyism would endure and to escape the Cold War trap he lived in Brazil, Israel, and the UK, eventually settling in London at Birkbeck College. In the meantime he met and married Sarah Woolfson, who became his life-long companion. We open this chapter with a section dedicated to Bohm's experience in Brazil. Then we break the chronological sequence to bring together in one place all the reaction to Bohm's causal interpretation; thus we dedicate Sects. 4.2–4.5 to the critics, supporters, mixed reactions and finally the Marxist milieu, successively. Then I resume the chronological thread. Section 4.6 is dedicated to the circumstances of application for Brazilian citizenship. Sections 4.7 and 4.8 are dedicated to his stay in Israel and the beginnings of his stay in the UK, in Bristol.

4.1 The Brazilian Experience: Expectation Frustrated and Passport Confiscated[1]

Bohm was dissatisfied not only with the political climate in the US but also with social and cultural life. "In a way I was looking forward to going to Brazil because I didn't like this whole atmosphere, and not merely the political atmosphere, but a lot

---

[1]Much of the material used in Sects. 4.1–4.4 was drawn, in many places expanded and/or rewritten, from my the Quantum Dissidents (Freire Junior 2015, 35–48).

© Springer Nature Switzerland AG 2019
O. Freire Junior, *David Bohm*, Springer Biographies,
https://doi.org/10.1007/978-3-030-22715-9_4

of the rest of the atmosphere. It was tied up also with the attitude to science and so on". Brazil, for Bohm, was a country at an earlier stage, as he recalled years later: "I was hoping that maybe their people would be a bit more old fashioned and would not have got caught up in this yet". As soon as he arrived, he wrote optimistically to Einstein, "The university is rather disorganized, but this will cause no trouble in the study of theoretical physics. There are several good students here, with whom it will be good to work". For Hanna Loewy and Melba Phillips, he illustrated the relaxed social manners in Rio de Janeiro telling the following anecdote about his arrival at the airport, "a cute little girl (about 18) came to take me to the waiting room. She asked me if I spoke Portuguese and I said "Not a word". She said "Surely you must know some words. [For example] the word for love—amor". Later, however, Bohm expressed his considerable dissatisfaction: "The country here is very poor and not as advanced technically as the U.S., nor is it as clean". "I am afraid that Brazil and I can never agree". "Brazil is an extremely backward and primitive country". His letters to Melba Phillips, Miriam Yevick, and Einstein are full of complaints about the state of the university, the fights in the physics department, the corruption in the government, the quality of the students, the weather, and the quality of the food.[2]

One month after his arrival Bohm suffered another blow from McCarthyism, demonstrating how its tentacles could cross borders and reach him far from the US. American officials in São Paulo confiscated his passport and told him that he could only retrieve it to return to his native country. As we know now, the US officials in São Paulo were following orders from the American State Department. Indeed, the FBI records say that "referenced report also reflects the State Department on October 2, 1951, requested the American Consul at Sao Paulo, to take up the subject's passport and hold it until the subject could arrange to return to the United States".[3] According to his report to Melba Phillips, "While I was in the physics dep't office, a representative of the Consulate came and told me that I should go to the consulate for 'registration and inspection of passports' as is required of all American citizens. When I went there, and registered, they informed me that they were keeping my passport, and that I could stay in Brazil, but that I could get my passport back only if I returned to the U.S. He would give me no more information". He went on to better explain his concerns: "What this means I don't know. At least, it means that they only want to be sure that I stay in Brazil, but at worst, it may mean that something more serious is cooking. Frankly, I am worried. I would appreciate it if you would watch the papers to see whether anything is developing in connection with the 'Weinberg case'." This profoundly changed Bohm's destiny and morale. He wrote to Einstein, "Now what alarms me about this is that I do not know what it means. The best

[2]Interview of David Bohm by Maurice Wilkins on 1986 September 25, Niels Bohr Library and Archives, American Institute of Physics, College Park, MD USA, www.aip.org/history-programs/niels-bohr-library/oral-histories/32977-4. David Bohm to Albert Einstein, Nov 1951, C.10–11, Bohm Papers. Both letters to Hanna Loewy and Melba Phillips were sent shortly after his arrival in Brazil on October 10, 1951. For these letters, see Talbot (2017, 116 and 133). "I am afraid that Brazil and I can never agree," Bohm to Loewy, October 6, 1953 (Talbot 2017, 125). All the letters to Hanna Loewy, Miriam Yevick, and Melba Phillips are transcribed and reprinted in (Talbot 2017).
[3]David Bohm FBI File, p. 152. I am thankful to the physicist and science writer Adam Becker for sharing with me the file he obtained thorough a FOIA request.

possible interpretation is that they simply do not want me to leave Brazil, and the worst is that they are planning to carry me back because perhaps they are reopening this whole dirty business again. The uncertainty is certainly very disturbing, as it makes planning for the next few years very difficult". Bohm's stay in Brazil, without a passport, changed his mood; he wrote to Melba Phillips: "Ever since I lost the passport, I have been depressed and uneasy, particularly since I was counting very much on [a] trip to Europe as an antidote to all the problems that I have mentioned".[4]

Bohm's mood oscillated depending on several factors, including the reception of his ideas among his fellow physicists. In addition, his hopes were not modest. "If I can succeed in my general plan, physics can be put back on a basis much nearer to common sense than it has been for a long time". Once he wrote, "I gave two talks on the subject here, and aroused considerable enthusiasm among people like Tiomno, Schutzer, and Leal-Ferreira, who are assistants ... Tiomno has been trying to extend the results to the Dirac equation, and has shown some analogy with Einstein's field equations". And then, "I am becoming discouraged also because I lack contact with other people, and feel that there is a general lack of interest in new ideas among physicists throughout the world".[5]

As we will see, in Brazil Bohm continued to work consistently on the causal interpretation, kept in contact with colleagues abroad, discussed his proposal with visitors from Europe and the United States, profited from collaboration with Brazilian physicists, and published results on the causal interpretation. Thus Bohm's activities in Brazil did not reflect the pessimistic views he expressed in some of the letters he wrote at the time. These letters tell us more about his mood, personality, and the context. That context was conditioned by political insecurity and by the adverse reception of his proposal among his fellow physicists, a subject I will consider in the following section. Bohm would have faced the many obstacles that he faced in Brazil elsewhere in working on a causal interpretation. Furthermore, Bohm's double identity as a Marxist and a Jew was not an issue for him in Brazil; on the contrary, it probably garnered him support. Brazil had been a *terre d'accueil* for Jews since the beginning of the 20th century and following the participation of the country in World War II with the Allies, the dictatorship called *Estado Novo* (1937–1945) ceded to a democratic regime. While political liberties were limited from 1945 to 1964, Communists could continue to play a role in Brazilian life. Examples are the writers Jorge Amado and Graciliano Ramos, the painter Cândido Portinari, the historian Caio Prado Jr., the physicist Mário Schönberg, and the architect Oscar Niemeyer.[6]

Moreover, Bohm arrived in Brazil at a propitious time for Brazilian physics. César Lattes had participated in the discovery of cosmic-ray pions in 1947 in the

---

[4]Bohm to M. Phillips, Dec 1951 (Talbot 2017, 139). Bohm to Einstein, Dec 1951, Bohm Papers (C.10–11). Bohm to M. Phillips, Dec 1951 (Talbot 2017, 141)

[5]Bohm to Melba Phillips, Nov 1951 and June 28, 1952 (Talbot 2017, 136 and 153). Bohm to Hanna Loewy, 6 Oct 1953 (Talbot 2017, 125).

[6]For more details on Bohm's stay in Brazil, see (Freire Jr. 2005, pp. 4–7 and 10–19). On Jews in Brazil, see (Rattner 1977); on Brazilian communist intellectuals, see (Rodrigues 1996, p. 412). During the 1930s, however, there were some obstacles for Jews in Brazil, see (Saidel and Plonski 1994).

UK, and in 1948 in the detection of artificially produced pions at Berkeley. These achievements resonated in Brazil, especially after the role of science in the war and the production of the first atomic bomb. An alliance among scientists, the military, businessmen, and politicians was developed so as to strengthen physics in Brazil. This led to the creation of the Centro Brasileiro de Pesquisas Físicas [CBPF] in Rio de Janeiro and, in the same year as Bohm arrived in Brazil, to the creation of the first federal agency exclusively dedicated to funding scientific research, CNPq. Bohm received several grants from CNPq to develop the causal interpretation. Visits to Brazil by Ralph Schiller and Mario Bunge, both his guests, and visits by Jean-Pierre Vigier and Léon Rosenfeld were afforded by this agency. Most of the money Bohm received went into research on cosmic rays, a field under Bohm's responsibility at USP. Nevertheless, the board of CNPq explicitly supported the development of the causal interpretation. An indication of the interest of CNPq in the research appears in the report by Joquim Costa Ribeiro, physicist and the Scientific Director of the agency, on Bohm's application for funds for Vigier's trip[7]:

> I call the attention of the Board to the interest of this subject. Prof. Bohm is today on the agenda of theoretical physics at an international level owing to his theory, which is a little revolutionary because it intends to restore to quantum mechanics the principle of determinism, which seems, in a certain way, to have been shaken by Heisenberg's principle. Prof. Bohm seems to have found one solution to this difficulty of modern physics, trying to reconcile quantum mechanics with the rigid determinism of classical physics. I am not speaking in detailed technical terms, but summarizing the issue. Bohm's theory has given rise to a great debate in Europe and United States, and Prof. Vigier has expressed his willingness to come to Brazil, mainly to meet the team of theoretical physicists and discuss the problem here. This seems to me to be a very prestigious thing for Brazil and our scientific community.

Bohm thus fought the most important intellectual battle of his life, the defense of the causal interpretation, in this context of different and even conflicting trends. He was anxious about the political situation in the US and concerned about Brazilian life, but he also had some support from Brazilian colleagues and institutions. Anxiety was also created by the criticism his causal interpretation was receiving and support from colleagues who supported it.

The research on the causal interpretation was not Bohm's only task in Brazil. In addition to his teaching duties, as the professor of theoretical physics at the Universidade de São Paulo, he also took care of the activities, experiments and theory of cosmic rays, a domain where Brazilian physics had excelled. He supported the activities of young physicists, such as Andréa Wataghin, George Schwachheim, and Hans Albert Meyer (Jean). He also supervised the activities of these physicists at the Chacaltaya Laboratory, in Bolivia, and was the responsible for bringing the senior physicist Kurt Sitte from the Syracuse University in the U.S to Brazil.[8]

---

[7]On Lattes' cosmic ray work, see (Vieira and Videira 2014). On Brazilian physics in the early 1950s, see (Brownell 1952) and (Andrade 1999). Costa Ribeiro's report is in Arquivos do CNPq (Records of the Conselho Diretor, 139th meeting, 25 Feb 1953), Museu de Astronomia, Rio de Janeiro.

[8]See Bohm's administrative letters at the Arquivo do Departamento de Física da FFCL-USP, Brazil, available on-line at: http://acervo.if.usp.br/. Accessed on 31 Jan 2019.

Still in Brazil, in the first years, Bohm also finished the work on plasma he had been doing with Pines when he left the US. He also dedicated time to understanding the foundations of statistical treatments which were part of the statistical mechanics theory. This work was done in intense interaction with the mathematician Miriam Yevick and with the Brazilian physicist Walter Schützer, whom Bohm had met while at Princeton. The physicist and historian of science Chris Talbot has analyzed the originality of these works and has shown that Bohm was able to make the first examination of deterministic chaos in physics after World War II. Bohm's interest in the subject was both scientific and philosophical. In particular, these works, together with his philosophical studies on determinism and chance in physics and on dialectic, nurtured his reflections on causality and chance which would appear in the book *Causality and Chance in Modern Physics*, published in 1957.[9]

In the following Sects. 4.2–4.4, I break the chronological order to gather the attitudes towards the causal interpretation in one place, in the 1950s, among the critics, the supporters, the mixed reactions, and the Marxist milieu (Figs. 4.1, 4.2 and 4.3).

4.2   The Causal Interpretation Campaign: The Critics

Bohm's approach to quantum mechanics did not pass unnoticed, as revealed by research into archives containing correspondence and papers from the early 1950s. As a matter of fact, most of the physicists who reacted to the causal interpretation were downright hostile to it, while a few of them became strong supporters, and a number of others had mixed reactions. Charting the initial reception of the causal interpretation is useful to shed light on Bohm's life but it is also illuminating of the dominant climate at the time towards research on the foundations of quantum physics. Wolfgang Pauli and Léon Rosenfeld were the first to react, Pauli even while the papers were in draft, as we saw in the previous chapter. Pauli concentrated on the physical and epistemological aspects, while Rosenfeld on the philosophical and ideological ones. As Rosenfeld explained his strategy to Pauli, "My own contribution to the anniversary volume [for de Broglie] has a different character. I deliberately put the discussion on the philosophical ground, because it seems to me that the root of evil is there rather than in physics". Let us first examine Pauli's reaction.[10]

After Bohm's papers appeared in print, Pauli advanced new criticisms, which Bohm knew of before their publication through letters from Guido Beck, who was in Rio de Janeiro, and Évry Schatzman, from Paris. Bohm was astonished: "I am surprised that Pauli has had the nerve to publicly come out in favor of such nonsense … I certainly hope that he publishes his stuff, as it is so full of inconsistencies and errors that I can attack him from several different directions at once". Pauli

---

[9]The work with Schützer led to the paper Bohm and Schützer (1955). Discussions with Miriam Yevick appear in his letters to her in Talbot (2017). The analysis of Bohm's work on statistical mechanics, by Chris Talbot, is in Talbot (2017, 49–61). Bohm's book is Bohm (1957).

[10]Léon Rosenfeld to Wolfgang Pauli, 20 Mar 1952, in (Pauli and Meyenn 1996).

Fig. 4.1 Brazilian physicists with Japanese physicist Hideki Yukawa at Princeton in 1949. Standing, from the right to the left: Walter Schützer, Hideki Yukawa, César Lattes, and Yukawa. Crouching: Jayme Tiomno, José Leite Lopes, and Hervásio de Carvalho. Tiomno invited Bohm to go to São Paulo when Princeton did not renew Bohm's contract. *Source* Archives of Centro Brasileiro de Pesquisas Físicas

Fig. 4.2 Jean-Pierre Vigier, the French physicist who became the main collaborator of David Bohm in the 1950s, in around 1947. This picture was obtained from Virgile Besson's doctoral dissertation, Besson (2018). *Credits* National Archives, UK, KV—Records of the Security Service, Reference: KV 2/1622

Fig. 4.3 Jean-Pierre Vigier, around 1992. *Credits* Courtesy of Joost Kircz

had criticized the causal interpretation for not preserving the symmetry between position and momentum representations, expressed in the standard formalism by the possibility of changing basis in the vector space through unitary transformations. He had also complained that Bohm's approach had borrowed the meaning of Ψ from quantum theory. In a letter to Markus Fierz, Pauli raised the stakes on the philosophical grounds criticizing the expectations of recovery of determinism in physics. He observed that Catholics and Communists depended on determinism to buttress their eschatological faiths, the former in the heaven to come, the latter in paradise on earth. These references were implicitly directed at Louis de Broglie on the one hand, and at Bohm and Vigier on the other. Pauli also warned his old friend Giuseppe Occhialini about "Bohm in São Paulo and his 'causal' quantum theory". Occhialini had worked in Brazil at USP during the 1930s and continued scientific collaboration there after the war. Pauli's substantive and persistent attack on Bohm's approach was based on two issues: Since it does not have "any effects on observable phenomena, neither directly nor indirectly ... the artificial asymmetry introduced in the treatment of the two variables of a canonically conjugated pair characterizes this form of theory as artificial metaphysics". And yet, "[if the] new parameters could give rise to empirically visible effects ... they will be in disagreement with the general character of our experiences, [and] in this case this type of theory loses its physical sense". Apparently, this criticism of Pauli went down well among physicists. "Incidentally, Pauli has come up with an idea (in the presentation volume for de Broglie's 50th birthday) which slays Bohm not only philosophically but physically as well," wrote Max Born to Einstein.[11]

Among the physicists who supported the complementarity view Rosenfeld played a singular role as a vocal and harsh critic of the causal interpretation. This role should be framed, however, considering the following issues. While he had been Niels Bohr's closest assistant for epistemological matters, as an adept of Marxism he saw the battle against the causal interpretation as part of the defense of what he considered to be the right relationship between Marxism and science. Indeed Rosenfeld was sensitive to criticisms against complementarity coming from the Marxist camp, particularly from the Soviets, even before the appearance of the causal interpretation, as we saw in the previous chapter. Rosenfeld went so far as to deny the very existence of a controversy on the interpretation of quantum physics, writing to Bohm, "I certainly shall not enter into any controversy with you or anybody else on the subject of complementarity, for the simple reason that there is not the slightest controversial point about it". For Rosenfeld, complementarity was both a direct result of experience and an essential part of quantum theory. Since complementarity implied the abandonment of determinism, as it precludes the simultaneous definition of position and momentum, which is the basis of mechanical determinism, Rosenfeld saw the causal interpretation as a metaphysical regression, writing, "determinism has not escaped this fate [becoming an obstacle to progress]; the physicist who still clings to it, who shuts his eyes to the evidence of complementarity, exchanges (whether he likes it or not) the rational attitude of the scientist for that of the metaphysician". Every good Marxist should understand that. "The latter [metaphysician], as Engels aptly describes him, considers things 'in isolation, the one after the other and the one without the other,' as if they were 'fixed, rigid, given once for all'". Rosenfeld believed that complementarity was a dialectical achievement that had to be defended not only against Bohm's criticisms but also against Soviet critics who blamed it for introducing idealism into physics. Rosenfeld's brand of Marxism was Western Marxism rather than the Soviet variety, to use the terms used by Perry Anderson. Thus Rosenfeld was orthodox in quantum mechanics and heterodox in Marxism.[12]

Orthodoxy, however, had its own internal tensions about how to express the meaning of the complementarity principle and Rosenfeld needed to deal with this. In his first publication on the dispute, a paper, in French, dedicated to the 50th anniversary of Louis de Broglie, Rosenfeld (1953) emphasized the idea of complementarity

---

[11]Bohm to Beck [w/d], Guido Beck Papers, Centro Brasileiro de Pesquisas Físicas, Rio de Janeiro. Évry Schatzman to David Bohm, May 15, 1952 (Besson 2018, 328). Beck had reported to Bohm the content of Pauli's seminar in Paris, in 1952. The criticisms were published in Pauli's contribution to the Louis de Broglie Festschrift, see (Pauli 1953a, b). Pauli to Markus Fierz, 6 Jan 1952, in (Pauli and Meyenn 1996, pp. 499–502); Pauli to Giuseppe Occhialini, [1951–1952]. Archivio Occhialini 5.1.14, Università degli studi, Milan. Max Born to Einstein, 26 Nov 1953, in (Einstein et al. 1971).

[12]For Rosenfeld's biography, see (Jacobsen 2012). Léon Rosenfeld to David Bohm, 30 May 1952, Léon Rosenfeld Papers, Niels Bohr Archive, Copenhagen (hereafter RP). For Western Marxism, see (Anderson 1976).

resulting from experience. Later, reacting to criticisms from Max Born, he polished his statement for the English publication of the same paper, changing "La relation de complémentarité comme donné de l'expérience" to "Complementarity and experience". And added, "but in any case the relation of complementarity is the first example of a precise dialectical scheme, whose formal structure has been accurately analysed by the logicians".[13]

Rosenfeld mobilized colleagues to take up the fight against the causal interpretation. He appealed to his professional connections as well as companions sharing ideological ties with Marxism. He urged Frédéric Joliot-Curie—a Nobel prize winner and member of the French Communist Party—to oppose French Marxist critics of complementarity. "I believe it is my duty to let you know about a situation I consider very serious which is happening close to you. It concerns your promising disciples Vigier, Schatzman, Vassails and *tutti quanti*, all intelligent young people and willing to do the best. Unfortunately, at this moment, they are suffering from grave illness. They are thinking it is absolutely necessary to kill complementarity and save determinism". Rosenfeld did not succeed as Joliot diplomatically chose to keep his distance from the battle, answering to Rosenfeld: "I agree with your concerns related the major principles of modern physics and I also agree on the need of a deep and precise understanding of these principles before going to discussions calling oversimplified citations, which may not correspond to the aims of their authors.[14]

Rosenfeld was more successful in the UK, where he taught in Manchester from the end of the war, than he was in France. He advised Pauline Yates—Secretary of the "Society for cultural relations between the peoples of the British Commonwealth and the USSR"—to withdraw her translation of a paper by Yakov Ilich Frenkel critical of complementarity from *Nature*; asked *Nature* not to publish a paper by Bohm entitled "A causal and continuous interpretation of the quantum theory;" and advised publishers not to translate one of de Broglie's books dedicated to the causal interpretation into English.[15]

Rosenfeld's correspondence shows that his campaign had wide support, as testified by Denis Gabor, "I was much amused by the onslaught on David Bohm, with whom I had a long discussion on this subject in New York, in Sept. 51. Half a dozen of the most eminent scientists have got their knife into him. Great honour for somebody

---

[13]Rosenfeld (1953). On Born's criticism, see (Freire Jr. and Lehner 2010).

[14]Léon Rosenfeld to Frédéric Joliot-Curie, 6 Apr 1952; Joliot to Rosenfeld, 21 Apr 1952. RP. See also (Pinault 2000, p. 508).

[15]Pauline Yates to Léon Rosenfeld, 7 Feb 1952, 19 Feb 1952, RP. Rosenfeld succeeded, "the editors stopped work on this article". The paper was submitted to *Nature* by Harrie S.W. Massey, with whom Bohm had worked at Berkeley, as we have seen in Chap. 2. *Nature*'s editors to Léon Rosenfeld, 11 Mar 1952, RP. "Also I sent a brief article to Massey with the suggestion that he publish it in *Nature*". David Bohm to Miriam Yevick, [November 23, 1951], (Talbot 2017, 204). Bohm did not keep a copy of the unpublished paper, but there is a copy of it in Louis de Broglie Papers, Archives de l'Académie des sciences, Paris. Léon Rosenfeld, "Report on L. de Broglie, La théorie de la mesure en mécanique ondulatoire". n.d. RP. The book Rosenfeld advised against translating was (Broglie 1957).

so young". Positive letters came from Abraham Pais, Robert Cohen, Vladimir Fock, Jean-Louis Destouches, Robert Havemann, and Adolf Grünbaum. Pais, who had been a student of Rosenfeld in Utrecht, wrote, "I find your piece about complementarity interesting and good … I could not get very excited about Bohm. Of course it doesn't do any good, but (with the exception of Parisian reactions) it also doesn't do any harm. I find that Bohm wastes his energy and that it will harm him personally a lot because he is moving into the wrong direction—but he needs to realize this himself, he is a difficult person". Cohen, a young Marxist physicist, wrote, "I turn to you because my own reaction to the Bohm thing and to the pilot wave revival has been quite negative, while yet I share Professor Einstein and others' uneasiness at the orthodox situation". Fock, who was the most influential and vocal supporter of complementarity in the USSR, wrote complaining that "Bohm-Vigier illness" was so widespread. Havemann, a German Communist physical chemist, sent him a paper on quantum complementarity, and Rosenfeld replied, "I read with great interest your paper and I am glad to see that our ideas are, in their essential aspects, in agreement".[16]

Rosenfeld's shots at the causal interpretation were not always well received. They also encountered resistance among his fellow physicists. Guido Beck and Eric Burhop, who had worked with Bohm at Berkeley during the war, took issue with Rosenfeld's rhetoric however, and Lancelot Whyte challenged him publicly over his review of Bohm's later book *Causality and chance in modern physics*. Guido Beck, once an assistant to Heisenberg who had fled from the Nazis to South America, did not share a belief in the causal interpretation, but defended Bohm against Léon Rosenfeld's criticisms and insisted Bohm should be encouraged to show what his approach could achieve. Rosenfeld was sensitive to Beck's remarks. In the English translation of the original French paper, Rosenfeld deleted the comparison which had been criticized by Beck. The original expression was: "on comprend que le pionnier s'avançant dans un territoire inconnu ne trouve pas d'emblée la bonne route; on comprend moins qu'un touriste s'égare encore après que ce territoire a été levé et cartographié au vingt-millième". Burhop, who had worked with Bohm on the plasma issue at Berkeley during the war, was at that time organizing a meeting among Rosenfeld and Marxist or left-wing physicists, such as John Bernal, Maurice Levy, Maurice Cornforth, and Cecil Powell to discuss Rosenfeld's article, also wrote: "Incidentally the only other comment I would offer on your article was I thought perhaps you were a little cruel to Bohm. Do you think you could spare the time to write to him? He is a young Marxist…being victimized for his political views in the U.S". Whyte, who was a Scottish engineer also interested in philosophy of science, considered the causal

---

[16]Denis Gabor to Léon Rosenfeld, 7 Jan 1953; Abraham Pais to Léon Rosenfeld, 15 May [1952]; Robert Cohen to Léon Rosenfeld, 31 Jul 1953; Robert Cohen to Léon Rosenfeld, 31 Jul 1953; Vladmir Fock to Léon Rosenfeld, 7 Apr 1956; all papers at *RP*. For Fock's criticism of Bohm's views, see (Fock 1957). For an analysis of Fock's philosophical views, see Martinez (2017). Jean-Louis Destouches to Léon Rosenfeld, 19 Dec 1951; Léon Rosenfeld to Robert Haveman, 7 Oct 1957; Haveman to Rosenfeld, 13 Sep 1957; Adolf Grünbaum to Léon Rosenfeld, 1 Feb 1956; 20 Apr 1957, 3 Oct 1957; Rosenfeld to Grünbaum, 14 Feb 1956; 21 May 1957; 11 Dec 1957. All letters are at RP. On Havemann, see (Hoffmann 1999).

interpretation anything but fleeting. Whyte considered Bohm's work comparable to Kepler's in mechanics, which surely was a compliment for a physicist.[17]

Rosenfeld went to Brazil to discuss the epistemological problems of quantum mechanics. He offered a course on classical statistical mechanics in Rio de Janeiro, published papers in Portuguese on the epistemological lessons of quantum mechanics, and gave a talk in São Paulo on complementarity. Bohm met him and reported on their exchange to Aage Bohr: "Prof. Rosenfeld visited Brazil recently, and we had a rather hot and extended discussion in São Paulo following a seminar that he gave on the foundations of the quantum theory. However, I think that we both learned something from the seminar. Rosenfeld admitted to me afterwards that he could at least see that my point of view was a possible one, although he personally did not like it". Bohm and Rosenfeld would meet each other again at a conference held in Bristol in 1957 and dedicated to foundational issues in quantum mechanics.[18]

Werner Heisenberg criticized Bohm's approach as "ideological" while Max Born initially was not impressed.[19] It was Rosenfeld who brought this interpretation to his attention, which led Born to criticize it. "I have already written my Guthrie Lecture in rough draft and have done there just what you suggest, namely, I have included the other party who prefer particles, like Bohm and the Russians which you quote (I cannot read Russian and I take it from your article.)" The common front against the causal interpretation hid disagreements, usually in private, over tactics. Rosenfeld publicly criticized Heisenberg of leaning towards idealism. Pauli and Born privately criticized Rosenfeld's mixture of Marxism with complementarity. As part of their debate, Max Born sent Rosenfeld a ten-page typed text arguing that dialectical materialism could not be corroborated by reference to just one achievement of contemporary science. Max Born abandoned the idea of publishing the text in the atmosphere of détente between West and East in the late 1950s. Acting as editor of a volume in honor of Bohr, Pauli prevented Rosenfeld, whom he labeled "$\sqrt{Bohr \times Trotzky}$," from adorning his paper with banalities on materialism.[20]

---

[17]Guido Beck to Léon Rosenfeld, 1 May 1952, RP. Rosenfeld to Beck, 9 Feb 1953; Bohm to Beck, 16 Sep 1952; 31 Dec 1952; 13 Apr 1953; 5 May 1953; 26 May 1953; Guido Beck Papers, Centro Brasileiro de Pesquisas Físicas, Rio de Janeiro. (Rosenfeld 1953). Eric Burhop to Léon Rosenfeld, 5 May 1952, RP. Lancelot Whyte to Léon Rosenfeld, 8 Apr 1958; 14 Mar 1958; 22 Mar 1958; 27 June 1958; Rosenfeld to Whyte, 17 Mar 1958, RP. Rosenfeld to Whyte, 28 May 1958, is in Lancelot L. Whyte Papers, Department of Special Collections, Boston University. The disputed papers were (Rosenfeld 1958) and (Whyte 1958).

[18]Bohm to Aage Bohr, 13 Oct 1953, ABP; (Rosenfeld 1954), (Rosenfeld 2005). For the Bristol conference's proceedings, see (Körner 1957). For an analysis of the Bristol conference, see Sect. 4.8 of this chapter and Kožnjak (2018).

[19]Heisenberg's criticism was published in the widely read and translated *Physics and Philosophy* (Heisenberg 1958). However, Heisenberg did not pursue the fight. In the late 1950s, "[he] had written more than enough on the subject and had, he said, 'nothing new to say'" (Carson 2010, p. 92).

[20]Born to Rosenfeld, 28 Jan 1953, RP. (Rosenfeld 1960, Rosenfeld 1970). Born's full text is in (Freire Jr. and Lehner 2010). Pauli to Heisenberg, 13 May 1954; Pauli to Rosenfeld, 28 Sep 1954, in (Pauli and Meyenn 1999, pp. 620–621 and 769).

While Rosenfeld, Pauli, and Heisenberg were the most active critics among the old guard who had created quantum physics (some mixed reactions will be analyzed later), among the younger generation criticism also predominated, albeit sometimes using different arguments. Bohm presented his approach at an international meeting held in Brazil and met open opposition to his ideas. As he wrote to Miriam Yevick, a colleague in the US:

> We had an international Congress of Physics … 8 physicists from the States (including Wigner, Rabi, Herb, Kerst, and others), 10 from Mexico, Argentina, and Bolivia, aside [a] few from Europe, were brought here by the UNESCO and by the Brazilian National Res. Council…. The Americans are clearly very competent in their own fields, but very naïve and reactionary in other fields…. I gave a talk on my hidden variables, but ran into much opposition, especially from Rabi. Most of it made no real sense.[21]

Isidor Isaac Rabi was an American physicist (born in Galicia) from Columbia University who had won the 1944 Physics Nobel Prize for his work using resonance for recording magnetic properties of nuclei. Bohm formulated Rabi's view thus: "As yet, your theory is just based on hopes, so why bother us with it until it produces results. The hidden variables are at present analogous to the 'angels' which people introduced in the Middle Ages to explain things".[22] Rabi's own statement of his criticism was,

> I do not see how the causal interpretation gives us any line to work on other than the use of the concepts of quantum theory. Every time a concept of quantum theory comes along, you can say yes, it would do the same thing as this in the causal interpretation. But I would like to see a situation where the thing turns around, when you predict something and we say, yes, the quantum theory can do it too.[23]

Bohm answered making a comparison with the debates on atomism in the 19th century, an analogy he had already used in his papers: "[E]xactly the same criticism that you are making was made against the atomic theory—that nobody had seen the atoms, nobody knew what they were like, and the deduction about them was gotten from the perfect gas law, which was already known". But Bohm faced tougher questions than his analogy suggested. How could the model be made relativistic? Anderson wanted to know how Bohm could recover the exclusion principle; A. Medina asked if Bohm's approach could "predict the existence of a spin of a particle as in field theory;" J. Leite Lopes and D. W. Kerst called for experiments that could decide between the interpretations; and M. Moshinsky asked whether there was a "reaction of the motion of the particle on the wave field". Bohm's answer to H. L. Anderson is interesting. He said that the causal interpretation only needed to reproduce the experimental predictions of quantum theory, not each one of its concepts. "All I wish to do is to obtain the same experimental results from this theory as are obtained from the usual theories, that is, it is not necessary for me to reproduce

---

[21] David Bohm to Melba Phillips, [July 1952] and Bohm to Miriam Yevick, [received 20 Aug 1952] (Talbot 2017, 155 and 271). I merged the two letters in my narrative.

[22] David Bohm to Melba Phillips, [July 1952] (Talbot 2017, 155).

[23] New research techniques in physics (1954, pp. 187–198).

every statement of the usual interpretation.... You may take the exclusion principle as a principle to explain these experiments [levels of energy]. But another principle would also explain them".[24]

Among other criticisms of the causal interpretation, it is interesting to note that of Mario Schönberg as it illustrates the complexity of the debate around the quantum controversy.[25] Bohm and Schönberg were both Jews and Communists (thus sharing the same Marxist views) but they failed to agree on one issue, the interpretation of quantum physics. Schönberg, a theoretical physicist, was working on the mathematical foundations of quantum theory and on the hydrodynamic model of quantum mechanics, a model close to that developed by Bohm and Vigier, as we will see later, but he opposed seeking a causal description in atomic phenomena. However, Schönberg (1954) exploited the physical implications of the quantum potential through hydrodynamic models. For instance, he showed that "the trajectories of the de Broglie-Bohm theory appear as trajectories of the mean motion of the turbulent medium". Despite their deep divergences, one of Schönberg's remarks was taken seriously by Bohm. Indeed, it was Schönberg who "first pointed Bohm in the direction of the philosopher G.W.F. Hegel, saying that Lenin had suggested that all good Communists read the German philosopher". This was an influence which would appear in Bohm's *Causality and Chance*, published in 1957. Unfortunately, at the time, Schönberg did not publish his views on the causal interpretation, but from Bohm's reaction to them one can infer how close to Rosenfeld he was on the subject at stake:[26]

> Schönberg is 100% against the causal interpretation, especially against the idea of trying to form a conceptual image of what is happening. He believes that the true dialectical method is to seek a new form of mathematics, the more "subtle" the better, and try to solve the crisis in physics in this way. As for explaining chance in terms of causality, he believes this to be "reactionary" and "undialectical". He believes instead that the dialectical approach is to assume "pure chance" which may propagate from level to level, but which is never explained in any way, except in terms of itself.

4.3 The Causal Interpretation Campaign: The Supporters

If the critics set the tone in the reception of Bohm's proposal, supporters were no less active, including attempts to further develop the initial papers. The most important adherents came from France with Louis de Broglie, who returned to his early ideas of a deterministic description of quantum systems, and Jean-Pierre Vigier, his

---

[24]All quotations are from New research techniques in physics (1954, ibid.).

[25]Other criticisms include Takabayasi (1952), Takabayasi (1953), Halpern (1952), Keller (1953), and Epstein (1953a, b). On Takabayasi's views, see (Besson 2018).

[26]For discussions between Bohm and Schönberg, see Peat (1997, pp. 155–157). David Bohm to Miriam Yevick, 8 Sep 1953 (Talbot 2017, 350). Schönberg (1954). For Schönberg's work on quantum mechanics and geometry, see also (Schönberg 1959). Schönberg's scientific works are collected and reprinted in (Schönberg and Hamburger 2009, Schönberg and Hamburger 2013).

young assistant. The importance of de Broglie's support may be inferred from the fact that Rosenfeld and Pauli chose to criticize Bohm's approach in their contributions to the de Broglie *Festschrift*, while the French Nobel Prize laureate was still cogitating the implications of Bohm's papers. Eventually de Broglie abandoned the complementarity view in the quest for a causal interpretation of quantum physics.

The influence of de Broglie's reconversion to his earlier ideas can also be seen in terms of the weight Rosenfeld attributed to it in a later letter to Niels Bohr: "This comedy of errors [the attempt to develop a 'theory of measurement' based on the 'causal interpretation' of quantum mechanics] would have passed unnoticed, as the minor incident in the course of scientific progress which it actually is, if it had not found powerful support in the person of L. de Broglie, who is now backing it with all his authority". In fact, de Broglie did not directly support Bohm's proposal, instead he pleaded for what he called the "double solution," which would remain as a mathematical suggestion and not a physical model for a causal interpretation. In 1953, through Vigier's visit to Bohm in Brazil, when their collaboration was already underway, de Broglie reminded Bohm of their differences: "You know our viewpoints are not entirely the same because I do not believe in the physical existence of the Ψ wave, which seems only to be the representation—rather subjective—of probabilities. By the way, when we have more than just one particle the Ψ wave must be represented in the configuration space with more than three dimensions and its non physical character appears then absolutely evident".[27] In fact, as remarked by historian Virgile Besson (2018), the collaboration among Bohm, de Broglie, Vigier and other physicists from the French team was, at least in its early stage and in its background, marked by tension and distrust. Thus Évry Schatzman, on May 15, 1952, while reporting on a seminar Pauli had given in Paris criticizing the causal interpretation wrote the following to Bohm: "L. de Broglie is a powerful. Getting subsides or financial help depends essentially of its good will. Unfortunately he is very sensitive, without practical sense, and we need its help for publications of papers on theoretical physics; on the other hand, we cannot trust him very much. One of its paper is made partly of ideas of Vigier expressed to him privately". Still, regarding the debates following Pauli's seminar: "The most remarkable was the speech of de Broglie. It can be summarized in the following way: if Bohm's theory is right, I said it before him, if Bohm's theory's wrong I also said it. But other people, less [subtle], I guess, were on the contrary, very much surprised that de Broglie took so firmly the defence of the pilot wave theory".[28]

Vigier brought momentum to the causal interpretation. He was influential among the French communists and in the Cold War times of the early 1950s he mobilized young Marxist physicists to work on the causal interpretation. With de Broglie and Vigier, the Institut Henri Poincaré became the world headquarters of the causal interpretation. A testimony from Jean-Louis Destouches reveals the isolation of complementarity point of view in the French milieu: "The young people received with

---

[27]Rosenfeld to Bohr, 21 Oct 1957, BSC, reel 31, AHQP. Louis de Broglie to Bohm, 29 March 1953, Louis de Broglie Papers, Box 7, Archives de l'Académie des sciences, Paris.

[28]The letter is in (Besson 2018, 328). The English is from the original letter, without corrections.

enthusiasm Bohm's work, which corresponds to the philosophical trends supporting their positions: Thomistic realism, Marxist determinism, Cartesian rationalism. I am almost the only one here to support Bohr's quantum interpretation".[29] Bohm also gathered support from the US, Argentina, and Brazil, through Hans Freistadt, Ralph Schiller, Mario Bunge, and Jayme Tiomno.[30]

Bohm considered the papers he wrote with Tiomno and Schiller and with Vigier to be the main achievements of the causal program in the early 1950s. With Vigier, Bohm answered Pauli's objection that he had included an arbitrary element in the causal interpretation, by using a ψ function that satisfied Schrödinger equation. Bohm had tried to solve the issue by himself without success, while De Broglie and Vigier were cognizant of the problem in 1952. In 1954, Bohm and Vigier were able to prove that under certain general conditions any function could become a solution of the Schrödinger equation. To achieve this, they used an analogy between Bohm's approach and the hydrodynamic model suggested by Erwin Madelung in 1926, which embedded microscopic quantum particles in a subquantum medium with random fluctuations. Thus, the "molecular chaos", an idea Bohm had abandoned after his discussions with Pauli, came back into his work with Vigier.[31]

Vigier's collaboration with Bohm was marked by their common Marxist background. In fact, in one of his first letters, Évry Schatzman, who was the intermediary for Bohm to contact Vigier, wrote to Bohm: "Any physical theory should be completely deterministic, because an affirmation of the dialectical materialism is that there is an objective reality and that this reality is cognizable, that we can built an image of that reality in our mind". Schatzman was far from modest about the work which was being done by Bohm and Vigier, comparing it to Marx's works: "We should be grateful to people like Vigier, like you, who have with tenacity devoted their efforts to the rebuilding of the quantum theory on its feet, just like the dialectic of Hegel, which had to be put back on its feet!" However, if the Marxist background was the cement, the collaboration between Bohm and Vigier blossomed in a fruitful scientific collaboration. We have already commented upon their first joint work, which introduced the model of a random sub-quantum level. Throughout the 1950s they tried, while without success, to obtain a relativistic generalization of the causal interpretation in order to reach similar predictions to those of the Dirac equation. "I have vowed to understand the Dirac equation, even if it kills me to do it. As I told Phil Smith, the day that we defeat the Dirac equation, we are going to have a special

---

[29]Jean-Louis Destouches to Léon Rosenfeld, 19 Dec 1951, RP.

[30]Freistadt worked both on the philosophical and technical aspects of the causal interpretation; on his activities in this subject in the context of American physics, see (Kaiser 2012, pp. 20–22). For Freistadt's works, see (Freistadt 1953, 1955, 1957). Schiller, Bunge, and Tiomno worked with Bohm in Brazil and their cases are discussed in this chapter.

[31]For the role Bohm played in these papers, see (Bohm 1980, pp. 114 and 118, notes 11 and 12); and (Bohm and Hiley 1993, p. 205). (Pauli 1953a, b). (Bohm 1953a, b); a simplified and shortened version of this paper was presented in (New research techniques in physics 1954, pp. 187–198). "This also was a one of the key issues Bohm had not dealt with in his 1952 papers". Jean-Pierre Vigier, interviewed by the author, 27 Jan 1992. (Bohm and Vigier 1954). For an analysis of the activities of the de Broglie-Vigier group, see (Vals 2012) and (Besson 2018).

victory party, with a case of champagne," he commented to Melba Phillips about the challenge. Later, he added: "The Dirac equation still doesn't come. By now I am convinced that a correct interpretation will require a treatment of second quantization (because of pair creation)". Still, "I got a lot of work done on the Dirac equation with Vigier in Paris. In fact, we now have a complete causal interpretation".[32] However, these results never reached a satisfactory level to be published.[33]

Still with Vigier, de Broglie, and other French physicists, in addition to Japanese physicists led by Takehiko Takabayasi, Bohm worked on an earlier promise. This was that his interpretation, conveniently modified, could be used in the domain of intra-nuclear particles, which he labeled in 1952 as the domain of 10^{-13} cm distances. They sought to classify the myriad of recently discovered intra-nuclear particles by representing them as extended bodies in space-time, relativistic rotators, and relating the number of degrees of freedom from these models to the quantum numbers attributed to the new particles. While this approach was neither a clear-cut extension of the 1952 model of electrons nor based on the requirement of causality, it was not strange to it. Indeed its philosophical assumptions were realism and the primacy of descriptions in the arena of space-time instead of abstract mathematical spaces. Before the appearance of the quark model, in the late 1950s, this was an exciting adventure in a new physical territory. In early 1960s however, this work receded when compared to other approaches which were based more on mathematical properties such as symmetry groups. All in all, it was a decade of intense collaborative work between Bohm and Vigier. Focusing on the activities of the French physicists, between 1951 and 1964, but taking into consideration their collaborations, the historian Virgile Besson described their story as a "biography of a minority research program".[34]

Jayme Tiomno had met Bohm at Princeton while he was doing his Ph.D. under John Wheeler on weak interactions. Ralph Schiller had worked on gravitation in his Ph.D. under the supervision of Peter Bergmann at Syracuse University and had gone to Brazil to be Bohm's research assistant. With Tiomno and Schiller, Bohm enlarged the scope of his model to include spin, although via analogy with Pauli's equation and not through a relativistic treatment of electrons. Tiomno, however, was not an adherent of the causal interpretation. He worked with Bohm looking for the consequences of extending Bohm's model to other fields of physics, but did not share its philosophical assumptions concerning causality. The Argentinian Mario Bunge, who had been supervised by Guido Beck at La Plata University, spent a year working with Bohm in Brazil, but nothing came of it. Bunge attacked the difficult problem of the "Bohmization" of relativistic quantum mechanics and the elimination of infinities in quantum electrodynamics. Bunge had studied physics in order to develop a better philosophy of the subject, later developing a successful career in the philosophy of science in Canada. In the mid-1960s, disenchanted with

---

[32]Bohm to Melba Phillip, [Fall 1952], 22 Nov 1954, 12 Oct 1955, (Talbot 2017, 157–174–180).
[33]See Talbot (2017, 47).
[34]Bohm and Vigier (1958); Bohm et al. (1960a, b); a review of the state of the art of this research is (de Broglie et al. 1963). Besson (2018).

the hidden variable interpretation, he gave up on it, accepted indeterminism as part of physics theories, and focused his criticisms of quantum mechanics on the role played by observation in the complementarity view.[35] Bohm and Bunge also had a common interest in philosophy of science, particularly on dialectics and nature. Indeed, before the arrival of Bunge in São Paulo, Bohm was reading Engels's *Dialectics of Nature* while Bunge had published on chance in nature in the Marxist journal *Science and Society*. Thus philosophy and physics were part of their shared agenda of interests.[36]

The collaboration between Bohm and Vigier, including the Vigier's trip to Brazil to work with Bohm was surrounded by an irony typical of the Cold War. Vigier went to Brazil on a leave of absence from his institution, the Institut Henri Poincaré, in Paris, and funded by the Brazilian agency CNPq, as we have already seen. Thus his stay in Brazil to collaborate with Bohm, endorsed as it was by Louis de Broglie, was a rather mundane trip for a physicist. Had Bohm however, remained in the U.S., Vigier might not have been able to visit and work with him. Vigier had made a name for himself in the Communist Party in France, as analyzed by Virgile Besson. However, as Jessica Wang has pointed out in writing about the "age of anxiety" in American history, "in addition to refusing passports to American scientists, the State Department also restricted the entry of foreign scientists with left-wing political ties into the United States ... Scientists from France, where the Left was particularly strong, had an especially hard time. As much as 70–80% of visa requests from French scientists were unduly delayed or refused". Collaboration with Vigier was of utmost importance for Bohm as the supporters who just applauded the causal interpretation on ideological grounds without trying to develop it did not help Bohm much. Apparently, this was the case of French astrophysicist, and Marxist, Évry Schatzman.[37] After all, the causal interpretation needed to overcome the technical challenges Bohm himself was trying to cope with.

4.4   The Causal Interpretation Campaign: Mixed Reactions

Not all reactions were clear-cut criticisms or support. The contributions of two people—Einstein and Feynman—were especially meaningful for Bohm. Einstein, the iconic critic of complementarity, had influenced Bohm while at Princeton to see quantum theory as an incomplete theory. On political grounds, Einstein was an enduring supporter of Bohm against McCarthyism. When the causal interpretation came out, however, he did not support it. "Have you noticed that Bohm believes (as de Broglie did, by the way, 25 years ago) that he is able to interpret the quantum theory in deterministic terms? That way seems too cheap to me," was his comment in a letter to Max

---

[35]Bohm and Schiller et al. (1955), Bohm and Schiller (1955). On Tiomno, see Freire Jr. (1999, p. 95). Mario Bunge to the author, 1 Nov 1996, and 12 Feb 1997.

[36]"I am reading "Diale'tica e Natureza", borrowed from the collections of Schönberg," Bohm to Miriam Yevick, 20 Nov [1951], in Talbot (2017, 200–203 [Letter 57]). Bunge (1951).

[37]Wang (1999, p. 279). Schatzman (1953). On Schatzman and the French milieu, see (Besson 2018).

Born. Moreover, he wrote a paper to a *Festschrift* for Max Born saying that Bohm's model led to the unacceptable consequence that particles in stationary states, such as an electron in a hydrogen atom, were at rest. Einstein may have used the opportunity to distance himself from the widespread opinion that he was stubbornly attached to determinism. "For the presentation volume to be dedicated to you, I have written a little nursery song about physics, which has startled Bohm and de Broglie a little. It is meant to demonstrate the indispensability of your statistical interpretation of quantum mechanics, which Schrödinger, too, has recently tried to avoid. […] This may well have been so contrived by that same 'non-dice-playing God' who has caused so much bitter resentment against me, not only amongst the quantum theoreticians but also among the faithful of the Church of the Atheists". Einstein, however, was kind enough to let Bohm read this paper before its publication and accepted Bohm's request to publish his reply in the same volume. Bohm showed that an adequate use of his model, including changes in the system due to measurements, could save it.[38]

Bohm's main hope for an ally among the foreign visitors he met in Brazil was Richard Feynman, who he had already met in the US and who spent his sabbatical year in 1951 at the Centro Brasileiro de Pesquisas Fisicas (CBPF) in Rio de Janeiro. Bohm liked Feynman's initial reaction, writing to his friend Hanna Loewy: "At the scientific conference at Belo Horizonte, I gave a talk on the quantum theory, which was well received. Feynman was convinced that it is a logical possibility, and that it may lead to something new". In a follow-up letter he was more detailed about his expectations of having Feynman adhering to the causal interpretation and leaving behind his work on renormalization procedures in quantum electrodynamics:

> Right now, I am in Rio giving a talk on the quantum theory. About the only person here who really understands is Feynman, and I am gradually winning him over. He already concedes that it is a logical possibility. Also, I am trying to get him out of his depressing trap of doing long and dreary calculations on a theory that is known to be of no use. Instead, maybe he can be gotten interested in speculating about new ideas, as he used to do, before Bethe and the rest of the calculators got hold of him.

This letter is evidence of how disconnected Bohm was at the time with the main themes of research on the physics agenda as he was criticizing as "dreary" the kind of calculations which were exciting not only Feynman and Hans Bethe, but almost all physicists involved with quantum field theories. This disconnection was not a consequence of a lack of knowledge. As we have seen in the previous chapter, Bohm had attended the Pocono and Shelter Island conferences and inspired by the mathematics Schwinger was using he used it in the treatment of plasma. Bohm had made a choice for plasma and later metal as a research program, which was later enhanced by his engagement with understanding quantum theory, first with his 1951 textbook and soon after with his papers on the causal interpretation. In hindsight we might even consider that Bohm had alienated himself from the physics mainstream. However, this was consequence of a choice as well as of the constraints engendered by this choice and the development of physics, a situation which reminds us of

---

[38]Einstein to Born, 12 May 1952 and 12 Oct 1953, (Einstein et al. 1971). (Einstein 1953), (Bohm 1953b). For Einstein's stances, see (Paty 1993, 1995).

"The Road Not Taken," the poem by Robert Frost. Bohm's hopes about Feynman were unfounded as "in his physics Feynman always stayed close to experiments and showed little interest in theories that could not be tested experimentally". The only reference Feynman made to hidden variables as a result of his Brazilian sabbatical was a possible avenue for the development of theoretical physics. Furthermore, it came out in a general paper published in a Brazilian science journal. That scarcely nourished Bohm's hopes of a stronger support to his interpretation.[39]

4.4.1 The Old Guard

From the old guard of quantum theory, let us now look at the cases of Niels Bohr, Erwin Schrödinger, and John von Neumann. Bohm particularly looked for reactions from Bohr and von Neumann, which is no surprise given that their views were the targets of his hidden variable interpretation. Bohm received the first report of Bohr's views through the American theoretical physicist Arthur Wightman, who was then in Copenhagen. As Bohm wrote to Melba Phillips: "the elder Bohr [Niels] didn't say much to Art[hur] Wightman, but told him he thought it 'very foolish.'" The distinction between the "two Bohrs" was particularly important as Bohm had met the younger, Aage Bohr, in the spring of 1948 while at Princeton,[40] and was pleased to discover that Aage Bohr was more sympathetic to the causal interpretation than his father, Niels Bohr, had been. As Bohm reported to Wightman, "I am glad that Aage Bohr admits its logical consistency".[41] Indeed, the younger Bohr [Aage] was more receptive to Bohm's proposal—"it would be nice to meet some time and discuss things, also the epistemological problems"—while he respected the value of the complementarity view: "there it seems to me that the very fact that one can give a logically consistent non-deterministic description of natural phenomena is a very great lesson which gives one a much freer way of thinking about things". The conversation continued and Bohm explained to Aage Bohr the two assumptions he considered to be "unnecessarily dogmatic" in the principle of complementarity: (1) "that the quantum of energy will remain indivisible and unanalyzable at all levels …", and (2) "that the statistical laws of quantum mechanics are final, in the sense that no deeper causal

---

[39]On Feynman's physics, see Schweber (2005). On Feynman in Brazil, see (Lopes 1990) and (Mehra 1994, pp. 333–342). David Bohm to Hanna Loewy, [Nov/Dec 1951] and Dec 1951, (Talbot 2017, 120–121), (Frost 1951, Feynman 1954). For the role played by Feynman, Bethe, and the renormalization calculations in physics at that time, see (Schweber 1994). However, Bohm only wrote such derisory words to Hanna Loewy, who was not trained in science. To his friends Melba Phillips and Miriam Yevick, physicist and mathematician, respectively, he was more cautious about Feynman's renormalization techniques, while calling them "resultlets". Bohm to Phillips, Nov 1951; Bohm to Yevick30 Apr 1952 and 9 Jan 1952 (Talbot 2017, 138, 257 and 308).

[40]David Bohm to Melba Phillips, Early 1952 (Talbot 2017, 147). Bohm wrote the same to Miriam Yevick, Feb 16 1952 (Talbot 2017, 247). Letter from Aage Bohr to the author, 17 Oct 1997.

[41]David Bohm to Arthur Wightman, [1953], Niels Bohr Archive, Copenhagen.

laws will ever be found …". [42] As for the elder Bohr, there was never any sign of empathy towards the causal interpretation, even after they had the opportunity for face to face conversation on Bohm's two visits to Copenhagen, in 1957 and 1958. As Bohm recorded five years later, Niels Bohr had "expressed especially strong doubts that such a theory [causal interpretation] could treat all significant aspects of the problem of *indivisibility* of the quantum of action".[43]

However, the main interest of Aage Bohr in the exchange with Bohm was not related to the epistemological issues in quantum mechanics, but to Bohm and David Pines' work on plasma, metals as electron gas, and collective variables. Aage Bohr was extending the collective variable approach to his own work on nuclear physics. He sent Bohm a preprint of a paper written with Ben Mottelson, and observed, "I would be also very interested in any comments from you on this, admittedly still rather primitive attempt of ours to develop a more comprehensive and self-consistent treatment of a many-body system such as the nucleus. In some ways, there are parallelities, I think, to your treatment of the electron gas, even though the forces and the geometry are quite different". Bohm, who was still in Brazil, was interested in Aage Bohr's work on nuclear physics, comparing it with results from the Van der Graaf accelerator being built in São Paulo. This would produce slow neutrons with very accurately determined energy.[44]

Thus in the late 1950s when Bohm was already in Israel and Pines was staying in Copenhagen, Bohm wrote to Aage Bohr. "I would very much like to spend [the summer] in Copenhagen and to work with Pines on plasma theory, on which subject both of us have interesting new ideas".[45] Bohm visited Copenhagen between 08 August and 29 September 1957 and then from 07 July 1958 to 13 September 1958. The influence of Pines and Bohm's plasma work on nuclear physics in Copenhagen was acknowledged, as we have seen in the previous chapter, by Ben Mottelson, the American physicist living in Copenhagen who went on, with Aage Bohr and Leo Rainwater, to win the 1975 Physics Nobel Prize for "the discovery of the connection between collective motion and particle motion in atomic nuclei and the development of the theory of the structure of the atomic nucleus based on this connection".[46]

---

[42] Aage Bohr to David Bohm, 3 Oct 1953; Bohm to Aage Bohr, 13 Oct 1953, emphasis in the originals, Aage Bohr Papers, Niels Bohr Archive, Copenhagen.

[43] Bohm (1962, 363).

[44] Aage Bohr to David Bohm, 3 Oct 1953; Bohm to Aage Bohr, 24 Sep 1953, ibid.

[45] Bohm to Aage Bohr, 18 Dec 1956, ibid. Aage Bohr replied, "I hope very much you can manage to come here next summer, when we also expect Pines to be here. We should, of course, be very pleased if you would tell us a little about plasma theory". Aage Bohr to Bohm, 26 Jan 1957, ibid. For the next summer, Aage mentioned they wanted to hear Bohm on superconductivity, reflecting the interest raised by the work of Bardeen and colleagues, Aage Bohr to Bohm, 25 Oct 1957, ibid.

[46] Visitors records, Niels Bohr Archive. "The Nobel Prize in Physics 1975", http://www.nobelprize.org/nobel_prizes/physics/laureates/1975/. "Ben R. Mottelson—Nobel Lecture", http://www.nobelprize.org/nobel_prizes/physics/laureates/1975/mottelson-lecture.html, on page 240. Both information accessed on 11 Jan 2014. (Bohr et al. 1958).

As for von Neumann, Bohm considered his reaction a little better than Bohr's. Bohm reported to Pauli that "it appears that von Neumann has agreed that my interpretation is logically consistent and leads to all of the results of the usual interpretation". And added the sources of such information: "(This I am told by some people.) Also, he came to a talk given by me and did not raise any objections". He sent more details to his friend Melba Phillips: "von Neumann thinks my work correct, and even 'elegant,' but he expects difficulties in extending it to spin". Von Neumann probably got interested in Bohm's work in the 1950s while revising the English translation of his *Mathematische Grundlagen der Quantenmechanik*, the German original edition had been published in 1932, in which his famous proof against the possibility of hidden variables appeared. To his publisher, he explained the difficulties, "the text had to be extensively rewritten, because a literal translation from German to English is entirely out of question in the field of this book. The subject-matter is partly physical-mathematical, partly, however, a very involved conceptual critique of the logical foundations of various disciplines". In a recent analysis, the philosopher Michael Stöltzner suggested that "von Neumann could accept Bohm's proposal as an interesting model, but not as a promising interpretation".[47]

As for Schrödinger, in spite of criticisms of the complementarity view, his insistence on the wave function ontology of the quantum world and absence of interest in the recovery of determinism hampered dialogue with those, such as Bohm, Vigier, and de Broglie, who worked with a world populated by particles in a deterministic framework.[48] However, Bohm was particularly annoyed by the hubristic reaction of Schrödinger. Unfortunately, we do not have Schrödinger's letter to Bohm but we can infer a little from Bohm's reaction in a letter to his friend Miriam Yevick, on February 16, 1952.[49] It also gives us a taste of the hardship of the campaign for the causal interpretation Bohm was embarking on:

> Schrödinger did not deign to write me himself, but he deigned to let his secretary tell me that His Eminence feels that it is irrelevant that mechanical models can be found for the quantum theory, since these models cannot include the mathematical transformation theory, which everyone knows is the real heart of quantum theory. Of course, his Eminence did not find it necessary to read my papers, where it is explicitly pointed out that my model not only explains the results of this transformation theory, but also points out the limitations of this theory to the special case where the equations are linear. In Portuguese, I would call Schrödinger "um burro", and leave it for you to guess the translation.

---

[47] von Neumann (1955). J. von Neumann's reaction is in David Bohm to Wolfgang Pauli, [Oct 1951], in (Pauli and Meyenn 1996, pp. 389–394); Bohm to Melba Phillips [Early 1952] (Talbot 2017, 147). John von Neumann to H. Cirker, [President of Dover Pub], 3 Oct 1949. John von Neumann Papers (Box 27, Folder 8), Library of Congress, Washington, DC. Stöltzner (1999).

[48] Schrödinger (1952 and 1953). On Schrödinger's philosophical views, see Michel Bitbol's comments in Schrödinger and Bitbol (1992, pp. 140–141) and Bitbol (1996). In private, Schrödinger kept at his fight against the complementarity view, as in this letter to Max Born, on October, 10, 1960: "The impudence with which you assert time and again that the Copenhagen interpretation is practically universally accepted, assert it without reservations, even before an audience of the laity—who are completely at your mercy—it's at the limit of the estimable ... Have you no anxiety about the verdict of history?" Moore (1989, p. 479)

[49] Bohm to Yevick, 16 Feb 1952 (Talbot 2017, 246).

4.5 The Causal Interpretation Campaign: The Marxist Milieu

Bohm was very sensitive to the reactions to his reinterpretation of quantum theory in terms of hidden variables. In particular, he paid attention to the way Marxists, physicists and philosophers, reacted to it, which is no surprise given Bohm's Marxist background. He made much of the French work, no doubt in part because of Vigier's Marxist engagement. "I have heard from someone that in a debate on causality given in Paris, when our friend Vigier got up to defend causality, he was strongly cheered by the audience (which contained a great many students). I will guess that many of the younger people in Europe recognize that the question of causality has important implications in politics, economics, sociology, etc.," he wrote to his friend, the mathematician Miriam Yevick. On the connection between the causal interpretation and Marxism, it appeared so obvious to Bohm that he complained when fellows like the American physicist Philip Morrison, whom he had met at his time at Berkeley, also a target for McCarthyism, did not support him. "This type of inconsistency in Phil [Morrison] disturbs me. He should be helping me, instead of raising irrelevant obstacles. Or at worst, he should do nothing. But to aid in the creation of confusion is something he ought not to do," he wrote to Melba Phillips on April, 1952. And he wondered why the causal interpretation had appeared in the West and not in the USSR and why Soviet physicists did not join him. "I ask myself the question 'Why in 25 years didn't someone in USSR find a materialist interpretation of quantum theory?' It wasn't really very hard. De Broglie + others suggested the 1st step in 1926, and the 2nd step (theory of measurements) is also not really difficult. But bad as conditions are in U.S. etc., the only people who have thus far had the idea are myself in U.S. and Vigier in France".[50]

There is no doubt that the influence of Marxism was effective in supporting the causal interpretation, especially among the French team led by Vigier, and that such support was influential to Bohm himself. However, it was weaker than Bohm had hoped for. The unfulfilled expectations were mainly related to the USSR, as evidenced in this long fragment of a letter he sent on March 18, 1955 to the American physicist Melba Phillips:[51]

> At times I feel discouraged about the state of the world. A thing that particularly strikes home to me is the report I got from Burhop (confirmed by others) on Russian physicists. Apparently, they are all busy on doing calculations on electrodynamics according to Feynman, Dyson, et al. Their orientation is determined strongly by the older men, such as Fock and Landau, who in addition to their training, are influenced by the fear of a sort of "Lysenko affair" in physics. The typical physicist appears to be uninterested in philosophical problems. He has

---

[50]David Bohm to Miriam Yevick, 5 Nov 1954 (Talbot 2017, 427–435, on p. 432). David Bohm to Melba Phillips, [April1952] (Talbot 2017, 150–153). Bohm to Yevick, 7 Jan 1952 (Talbot 2017, 225–230, on p. 230).

[51]Bohm to Melba Phillips, 18 March 1955 (Talbot 2017, 176–179). Andrew Cross (1991) saw Bohm's work as just a reflection of the ideological Marxist climate of the time; thus he missed the fact that the quantum controversy continued even when that climate faded. For the critique of this position, see (Freire Jr. 1992).

not thought much about problems such as the re-interpretation of qu. mechs., but tends to take the word of the "bigshots" that ideas on this such as mine are "mechanistic". Actually, the standard procedure is just to label such a point of view, and then most people accept the label without even bothering to read about such questions. There are some philosophers in Moscow who criticize the usual interpretation, but they haven't had much influence on the physicists. All in all, the situation in Soviet physics doesn't look very different from that in Western physics. It is disappointing that a society that is oriented in a new direction is still unable to have any great influence on the way in which people work and think. In both societies, one finds that the typical person finds it safer and more convenient to avoid "controversial" subjects and to become an expert in some narrow technical field. I must say that there is some justification in the effort to avoid another example of the extremes to which Lysenko went in criticizing backward trends in biology. But the reaction has been for people simply to avoid fundamental questions altogether, and to say that experiments will eventually indicate the new theories that are needed, while meanwhile, everyone works on the old ideas until this happy day will be reached.

What Bohm did not realize was that part of the support for complementarity as well as resistance to the causal interpretation was also based on commitment to Marxism. This was the case of Rosenfeld and Schönberg, as we have already seen. This was also the case of Vladimir Fock, who supported Bohr's views in the USSR basing his position on dialectical materialism. From 1957 on, after Stalin's death and the ideological thaw in the USSR, Fock would become an outspoken defender of Bohr's views. In addition, a number of Soviet physicists, such as Blokhintsev and Terletsky, while critics of complementarity, were not supporters of the causal interpretation either. Indeed, the former became a leader in the defense of the ensemble interpretation, which says quantum theory does not describe states of single systems but only an equally prepared ensemble of them.[52] The latter devoted his energies to attempts to include non-linearities in the standard quantum mechanics, an approach which resonated with de Broglie's proposal of a "double solution". Indeed, we may see in hindsight, the relationship between Marxism and the spectrum of stances in the quantum controversy was not one-to-one. Instead, Marxism influenced both critics as well as defenders of complementarity. This multi-sided relationship should be of no surprise as when speaking of Marxism in the 20th century it is better to use the plural Marxisms than the singular Marxism.[53]

---

[52]On the relation between Bohm and the philosophy of science in the Soviet Union, see Talbot (2017, 77–88). For a description of the ensemble interpretation, see Home and Whitaker (1992) and Kojevnikov (2011).

[53]On Marxism and the controversy over the interpretation of quantum theory, see (Freire Jr. 2011). See also (Graham 1987, pp. 320–353), on Fock and Blokhintsev; (Martinez 2017), on Fock; (Kuzemsky 2008), on Blokhintsev; (Pechenkin 2012), on the early ensemble interpretation in the USSR and in the US; (Forstner 2008), on Bohm; (Jacobsen 2007, Jacobsen 2012), on Rosenfeld; (Kojevnikov 2011), on ensembles; (Pechenkin 2013), on Mandelstam; (Kojevnikov 2004), on Soviet physics, and (Besson 2018), on Vigier.

4.6 The Price of Travelling Abroad, Brazilian Citizenship Gained and American Citizenship Lost

In early 1952, as part of the deployment of the campaign for the causal interpretation, David Bohm began to think of ways to get a passport to travel abroad in order to discuss his interpretation with fellow physicists. Following a contrived plan he was able to get a Brazilian passport, through obtaining Brazilian citizenship, and then left Brazil for a position in Haifa, Israel, from where he made regular trips to Paris, London, and Copenhagen. After about two years in Israel he left for a temporary position in Bristol, in the UK. The success of his plan came with a price, typical of the Cold War times: he lost his American citizenship.

Following the confiscation of his passport, Bohm began to consider applying for Brazilian citizenship. It would be a difficult choice, as he wrote to Hanna Loewy: "I have gotten a permanent identity card, which means that I can stay here indefinitely, even without a passport. Also, if I want, I can apply for citizenship. This would have some advantages; as with it, I could travel. But the disadvantage is that I could not return to the US, at least for a long time. For according to the McCarran act, they can exclude any non-citizen from the U.S. who in their opinion, was ever connected with Communism. So it's a tough decision, isn't it?"[54]. At the end of 1953, anxious because of the tension of the political times (a year before the Rosenberg couple, accused of espionage, had been executed in the U.S.), he began to making plans for England and Israel.[55] The latter seemed to be the most viable as he had a potential offer from Nathan Rosen, who was the head of the Physics Department at Technion in Haifa, and this gave him two different ways to obtain a passport, as he wrote to Lilly Kahler:[56]

> There are two possible ways of getting to Israel. One is to arrange the job offer and apply to the State Dep't for a passport on the basis of this offer. Perhaps the Israeli gov't would be willing to make some representation to the State Dep't.
>
> The second is to go to the Israeli embassy here and ask for a special document to travel to Israel. This method would imply that I intend to become a citizen there.

For both choices he needed Einstein's support, as he commented to Lilly Kahler asking her to intermediate in the subject with Einstein: "Einstein's help would be especially good in this connection. I'm not asking him to do anything against his principles but would he: (a) Write a letter of recommendation for me to get a job in a place like Haifa? (b) In case of necessity, write a letter to the Israeli embassy in Brazil, suggesting that they give me a special document to travel to Israel?"

---

[54]Bohm to Hanna Loewy, [December 1951] (Talbot 2017, 121).

[55]On the trial and execution of Julius and Ethel Rosenberg, including an update information about the case, see Schrecker (2002), particularly pp. 38–42.

[56]Bohm to Lilly Kahler, 3 Dec 1953 (Talbot 2017, 128).

Einstein was reluctant, writing to Bohm in early 1954, "to go there with the intention to leave on the first occasion would be regrettable". Despite this, Bohm decided to go, writing to him on February 3: "I have decided to go to Israel. This decision was precipitated by the receipt of an offer of a job in Haifa from Rosen [...] I have cited you as a possible recommendation, so you may be receiving a letter from them soon". He also promised to stay there for years to Einstein ("... do not plan to leave unless after several years of effort"), a promise he would not keep. In addition, Bohm was thinking of obtaining an Israeli passport, "I am informed that the Israeli Embassy in Brazil may issue a passport for me to go to Israel, if the Technion request it". Einstein then changed his mind and supported Bohm's plans.[57]

While the Haifa job was confirmed nothing concerning the Israeli passport materialized. In the end, he went back to the idea of applying for Brazilian citizenship. In early 1953, Bohm had estimated a time of one and a half years to obtain citizenship: "This I cannot do in less than about 18 months from now; it may take two years".[58] This time, however, the whole process was quicker than he had imagined as Bohm was helped by Brazilian scientists and politicians in a procedure typical of Brazilian culture: the use of good social connections to undercut administrative deadlines. Indeed, he applied for citizenship on 15 September 1954, received the presidential decree on 22 November 1954, and took the oath on 20 December 1954, thus it took three months in all. In early 1955 he left the country for Israel. The University of São Paulo took his departure as a leave of absence, unpaid, for one year, expecting him to come back, which did not happen. Only on March 1, 1956, was his contract eventually ended (Fig. 4.4).[59]

Getting Brazilian citizenship, however, was a fateful decision for Bohm the full consequences of which will appear later as it led to the loss of his American citizenship. As early as April 1955, Marc Severe, an official from the American consulate, required the Brazilian Police Department to give the US government information about the Brazilian nationality of Bohm. Brazilian authorities took time replying, but eventually they confirmed that Bohm had been granted Brazilian citizenship. Bohm lost his American citizenship on 5 December 1956.[60]

---

[57]Einstein to Bohm, 22 Jan 1954; Bohm to Einstein, 3 Feb 1954; Einstein to Bohm,10 Feb 1954; and Einstein to Nathan Rosen, 11 March 1954. The Albert Einstein Archives, The Hebrew University of Jerusalem. My thanks to Michel Paty and Amit Hagar for providing me with copies of these letters.

[58]Bohm to Lilly Kahler, 3 Dec 1953 (Talbot 2017, 128).

[59]According to physicist José Leite Lopes [Interview with Ana Maria Ribeiro de Andrade, 18 March 2003], Brazilian physicists had asked João Alberto Lins de Barros, a very influential politician and supporter of Brazilian physics, to accelerate Bohm's Brazilian citizenship application. File 40.135/54. Archives of the "Instituto de Identificação Ricardo Gumbleton Daunt", SSP—Polícia Civil, São Paulo. David Bohm's administrative records are from Arquivo da Faculdade de Filosofia, Ciências e Letras, Universidade de São Paulo, Brazil.

[60]Idem.

Fig. 4.4 The physics department at the University of São Paulo where Bohm worked from late 1951 to early 1955. *Credits* Acervo CAPH/FFLCH/USP Imagens

4.7   Israeli Times and the Trauma of Rupture with Communism

Technion at Haifa was set up in the early 20th century, thus predating the State of Israel, It was at this institution where Bohm spent the years 1955–1956. He had been introduced to Nathan Rosen, the head of the physics department by David Fox, with whom Bohm had been acquainted since his days at Berkeley; Fox had also been a victim of McCarthyism and was then working at Technion.[61] There Bohm also met Kurt Sitte, a cosmic ray experimentalist who had worked with Bohm in São Paulo. Rosen had had interests in the foundations of quantum mechanics since his times in the US, and was the "R" in the famous EPR paper where Einstein, joined by Boris Podolsky and Rosen, had criticized the completeness of quantum mechanics. At Technion, Bohm was a dedicated lecturer, as he had been at Princeton and São Paulo. He had learned Portuguese in Brazil and gave classes in Portuguese, and now he learned Hebrew and in 1956 was already giving classes in this language. He was hired as a lecturer and soon was promoted to associate and then to professor. Bohm profited from his stay in Israel to freely travel to Europe with his Brazilian passport. The main result was the enhancement of the collaboration with Jean-Pierre Vigier and his team at the Institut Henri Poincaré in Paris, but there were also trips to London, Bristol, and Copenhagen. However, Bohm was not content in Israel, complaining about the weather, the organization of the university, and the lack of people to discuss his work. "It is evident to me now that the hot weather of Israel is bad

---

[61]"Dr. David Fox, 78, a Physicist Blacklisted in the McCarthy Era," The New York Times, 10 May 1999.

for me. It makes it hard for me to work and is conducive to intestinal disturbances. Anyway, I have pretty well decided to go to Europe next year, either England or France," were his words to Melba Phillips.[62]

However grim Bohm's recollections may have been, they did not express the scientific achievement he was able to produce there. He met two bright students, Gideon Carmi and Yakir Aharonov, who would follow him on the next step of his journey to the UK. Still at Technion, with Aharonov, Bohm produced a milestone paper in our understanding of the quantum. In their first joint work, they compared the ideal experiment suggested by Einstein, Podolsky, and Rosen in 1935 with real data using results from a previous experiment with photons produced at an annihilation of a positron-electron pair. This experiment had been performed by the Sino-American physicist C. S. Wu and I. Shaknov in the late 1940s with a different purpose.[63] Bohm and Aharonov translated the continuous variables original EPR into an experiment concerning spins of electrons, which Bohm had already done in his 1951 textbook *Quantum Theory*, and then into photon polarizations. Finally, they fitted this treatment to what was given in the Wu's experiment, that is, the "measuring the relative rate, E, of coincidences in the scattering of the two photons through some angle". All this is a standard quantum mechanics treatment. Furthermore, they modeled what one should expect in such an experiment if the quantum state of the pair of photons (an entangled state, we would call nowadays) were not valid for after some separation between the photons, a hypothesis which had been envisioned by Einstein and by Wendell H. Furry. In other words this hypothesis says that "the current formulation of the many-body problem in quantum mechanics may break down when particles are far enough apart". After calculation, they presented the results in Table 4.1.

Results in line A were calculated following quantum mechanics to the ideal EPR experiment. Lines B represent calculations under the assumption quantum mechanics would no longer be valid for systems far away from each other. Results from observations made by Wu and Sakhnov are in the last line. Experimental results thus corroborated quantum mechanics and invalidated the assumption that it would fail for distant systems. As important as the results themselves was the analysis they suggested for these data: "The results in Table 4.1 show that this experiment is explained adequately by the current quantum theory which implies distant correlations, of the type leading to the paradox of ERP, but not by any reasonable hypotheses implying a breakdown of the quantum theory that could avoid the paradox of ERP". In other words, Bohm and Aharonov are saying that quantum mechanics implies entanglement as a physical effect. This is precursor to Bell's theorem even if it is not as clean

---

[62] David Bohm to Melba Phillips, 12 Oct 1955, (Talbot 2017, 180). On his perception of Technion, see David Bohm, interviewed by Maurice Wilkins, session 6, 22 Dec 1986, pp. 65–66, Niels Bohr Library and Archives, American Institute of Physics, College Park, MD.

[63] Bohm and Aharonov (1957). All the citations are from Bohm and Aharonov (1957); the experiment is Wu and Shaknov (1950). Incidentally, I noticed that the Bohm and Aharonov paper is careless on two citations. They systematically refer to ERP instead of EPR, thus changing the name order between Podolsky and Rosen, which Bohm had already done in his 1951 book *Quantum Theory*, and they did not cite the theoretical investigation by M. H. L. Pryce and J. C. Ward, which anteceded Wu and Shaknov's experiment and is cited by them.

Table 4.1 Relative probability of coincidences in the Compton scattering of annihilation photons for the two geometries of Fig. 1.1

Hypothesis	Scattering probability $d\,\Sigma$ divided by $(r_0^4/8)(d\Omega)^2(k^4/k_0^4)$		Ratio $R = d\Sigma_\perp/d\Sigma_\parallel$	
	π_1 parallel π_2	π_1 perpendicular π_2	For ideal angle 82°	Average for experimental solid angle
A (standard)	$2\gamma(\gamma - 2\sin^2\theta)$	$(\gamma - 2\sin^2\theta)^2 + \gamma^2$	2.85	2.00
B_1 (product of states of opposite circular polarization)[a]	$d\Sigma$ unaffected by orientation of π_1 relative to π_2		1.00	1.00
B_2 (product of states of perpendicular linear polarization, randomized directions)	$(2\gamma^2 - 4\gamma\sin^2\theta + \sin^4\theta)$	$(2\gamma^2 - 4\gamma^2\sin^2\theta + 3\sin^4\theta)$	<2	~1.5
B_{inter} (intermediate case of elliptic polarization)	Intermediate between B_1 and B_2		<2	<1.5
Observation (reference 10)				2.04 ± 0.08

[a]Equal probability for $x_1 = \psi + (1)\psi - (2)$ and for $x_2 = \psi - (1)\psi + (2)$. The correct wave function according to standard quantum theory is $2^{-1/2}(x_1 - x_2)$

and neat as the theorem. It is a statement that quantum theory leads to entanglement, the quantum correlation of systems which are far away from each other, and that experiments corroborate this feature of quantum mechanics. Due to this, this work by Bohm and Aharonov, as well as Wu's experiment, were seen by some as the precursor of the work with Bell's theorem. The paper published by Bohm and Aharonov has garnered today more than 330 citations.[64]

This paper still presents unique features to single out Bohm's own evolving ideas on the interpretation of quantum mechanics. For the first time in print, the paper was submitted on May 10, 1957, Bohm no longer defended his original 1952 interpretation, acknowledging "It must be admitted, however, that this quantum potential seems rather artificial in form, besides being subject to the criticism of Sec. 4.1 that it implies instantaneous interactions between distant particles, so that it is not consistent with the theory of relativity". He went further to suggest a new interpretation,

---

[64]Duarte (2012), Maia Filho and Silva (2019). Bohm and Aharonov (1957). Citations from *Web of Science*, accessed on 13 Sept 2018.

based on the assumption of a "deeper subquantum-mechanical level" along the lines of the work he had done with Vigier and published in 1954.[65] According to Bohm and Aharonov, "The laws of this lower level are different from those of the quantum theory, but approach these latter laws as an approximation". While they announced that "explanations of this kind will be published later," such a work never came out.

A few weeks after Bohm arrived in Israel he met Sarah Woolfson and they began to date and married one year later, in 1955. David and Sarah would spend the rest of their days together. There is no doubt that Sarah was a calming influence on Bohm's anxieties. They had no children.

In the fall of 1956 Bohm and Sarah embarked on an extended trip to Europe. There Bohm would suffer two major blows. The first was the Suez crisis, or the second Arab-Israeli War, when Israel, followed by the United Kingdom and France invaded the Egypt. Bohm's immediate conclusion was that it would be unsafe to return to Israel. However, negotiations with British universities, which were speedy enough, conflicted with Bohm's commitments in Israel. Indeed, his contract at Technion was still valid and classes would be resumed soon. We only know what happened while they were in Europe from his letter to Melba Phillips when he had already returned to Haifa: "Well, we are now in Israel again. Whether it was a wise decision, we can't know. Many factors entered into it. First the English university people hinted to me that if I broke my contract with Haifa, I might find it difficult to get a job there. Later, they changed their mind, but when I heard of it we were all ready to go. Then, I feel that it is bad to break from a distance in this way". Bohm added that "classes are going on as normal: but I am already two weeks late". Indeed in the summer, on June 20, 1946, at Technion, the appointments committee had already decided on the promotion of Bohm to a full professorship.[66] Bohm's relationship with Technion however, had been irreversibly fractured. In the following year, 1957, he left his tenured position at Technion for a research fellowship at University of Bristol, appointed by Maurice Pryce, who he had met while at Princeton.

The second blow was his disillusion with Communism, following the revelations of Nikita Khrushchev, the Soviet leader, at the 20th Congress of the Soviet Communist Party, about the crimes in the Stalin-era, and the Soviet invasion of Hungary. Bohm's break with Communism happened while he was in Paris, alone, as Sarah had gone to the UK to visit her relatives. According to Bohm's later recollections, "This was a tremendous shock to me. I had always hoped that socialism would be a way to approach these problems that humanity was unable to solve". Bohm's experience while learning this news was witnessed in Paris by the physicist Jan Meyer, who reported the same as Bohm's later recollections—"I remember walking around Paris for hours on end"—and is well-recorded in two long letters to Melba Phillips. How

---

[65]Bohm and Vigier (1954).

[66]Bohm to Melba Phillips, [November 1956], (Talbot 2017, 193). On the appointments committee, see the minute of Technion—Committee for Academic Appointments, 20 June 1956. I am thankful to Amit Hagar for providing me with such a document and Elaine Ruth Fletcher and Michael Liss for the translation from Hebrew to English.

dramatic Bohm's involvement was with these critical events may be seen from the following fragments:[67]

> It is clear from the above that what is needed in the left-wing movement today is a certain measure of disengagement from Russia. Russia has made an enormous number of errors.... This raises the question of the probable future of the C.P.'s [Communist Parties] throughout the world.... As soon as a man opposed the direction of the C.P. he became a traitor, guilty of the most heinous crimes. Confessions were manufactured and extorted on a large scale. The truth had nothing to do with the case: what was published was only what would be convenient for the interests of the gov't. This was a direct perversion of the principle that dialectical materialism should be scientific and objective. Perhaps some people said that false confessions served the interests of a 'larger truth'. Similarly, Humanité [The French C.P. newspaper] still publishes lies about Hungary, quite cynically, since the truth is evident. It is clear also that the Russian gov't publishes whatever it thinks is convenient about world affairs. Perhaps they have already ceased to lie consciously, and they may be only deceiving themselves.

Bohm's rupture with Communism in 1956 was not an exceptional movement. Indeed, similar disenchantment spread around the globe. In the case of the American communists, with whom Bohm kept in touch even at a distance through correspondence, historian Maurice Isserman relates that "a full three quarters of the American Communist Party membership, people who had stayed with the movement in the worst years of McCarthyism, quit in the year or so after the events of 1956". D. Shannon, another historian, acknowledges the inaccuracy of the figures, "how many members left the party in the fall of 1956 will never be known since the Communists did not conduct a party registration until the spring of 1958". However, he identifies an estimate for the numbers: "yet party officials knew that hundreds were quietly leaving". Shannon's adjective, "quietly," may be illustrated with the case of Howard Fast, the acclaimed novelist author of *Spartacus*. According to Shannon, "Howard Fast, still in the emotional turmoil that characterized his column on the Khrushchev speech, left quietly without announcement. His leaving made no stir until New York newspapers discovered his resignation in February, 1957". David Bohm's move was similar, he "left quietly without announcement".[68]

While Bohm's long letters focus on a rupture with Stalin's legacy and practice and still consider prospects of a new left-wing worldwide movement, they indeed marked a rupture with Marxism as, from then on, there are no other manifestations from Bohm towards Marxism except for taking a critical distance from this worldview. In fact this

---

[67]For an account of those events, see (Gaddis 2005, pp. 83–194) and Westad (2017, pp. 198–206). Interview of David Bohm by Maurice Wilkins on 1986 December 22, AIP, www.aip.org/history-programs/niels-bohr-library/oral-histories/32977-6. Jan Meyer, conversation with Olival Freire, 30 January 1997; Bohm to Phillips, two letters, [not dated] and [November 1956], (Talbot 2017, 182–197). Bohm's rupture with Communism possibly strained his intellectual relationship with Melba Phillips, who did not follow his stands (from what we can infer from Bohm's letters) as there was no follow up of their correspondence. This rupture is also noted by Kojevnikov (2002, p. 191) and Peat (1997, p. 178). For an analysis of Bohm's position in relation to the stances of the Soviet Communist Party, see Talbot (2017, 67–76).

[68]On the case of the French communists, see Caute (1967) and Ory and Sirinelli (2004). Isserman (1980, 44). Shannon (1959, 317–318).

rupture resonated with the philosophical view he was shaping and the attenuation of defense of the causal interpretation, a subject we shall come back later. Thus by the end of 1956 Bohm's relationship with Marxism came to an end after about thirteen years. Commitment to Communism in the Cold War context in the US had cost him the right to live in his home country and would still fuel a lasting battle to recover his American citizenship. He lived the main political passions of his times and was a man trapped in the Cold War storm. And yet, his history, including adhesion to and a later break with Communism, was not exceptional, indeed it was typical of the generation of intellectuals in the mid-20th century, around World War II.[69]

4.8 Bristol, the Philosophical Stage, and Scientific Breakthrough

David Bohm arrived in Bristol for a four-year research fellowship. Bristol was then a thrilling center for physics under the leadership of Maurice Pryce, the head of the Department. "Pryce appointed David Bohm (1917–1994), who arrived in 1957 with his student Yakir Aharonov (b. 1932). Their discovery [...] was central to the formulation of modern gauge theories of fundamental interactions". These are the recollections of Michael Berry and Brian Pollard on Bohm's stay at Bristol in the late 1950s. The years in Bristol were fruitful for Bohm for distinct reasons. He was ready to enter the philosophical stage discussing ideas not only with physicists but also with philosophers; he and Aharonov got a result—the Aharonov-Bohm effect—which would be very influential in theoretical physics; and his years there paved the way for his appointment to a permanent position at Birkbeck College, in London, in 1960. Despite of all these achievements, Bohm did not have good recollections of his times in Bristol. He complained of the hierarchical and competitive ambiance at the physics department; he thought Pryce was jealous of his projection, and he thought philosophers were narrow-minded in their appreciation of his major scientific-philosophical work, the book *Causality and Chance in Modern Physics*. "First of all, there was a tremendous status consciousness there. When Pryce was in America, he seemed very open and very democratic or whatever you want to say, but in Bristol he was very conscious of his status as head of the department. There was a sense of hierarchy which showed in subtle ways of difference and all that. [...] Not only Pryce, but each other. There was a sort of general sense of people watching each other". These were Bohm's recollections, in 1986, when interviewed by Maurice Wilkins. As for the reception of the book, "[it] was published just about the time I got to England. It got a number of reviews, which were very good, but the philosophical fraternity paid very little attention to it because I'm not one of them".[70]

---

[69](Ory and Sirinelli 2004), (Hobsbawm 2011, Chaps. 11 and 14), (Caute 1967).

[70]Berry and Pollard (2008). Interview of David Bohm by Maurice Wilkins on 1986 October 3, and December, 22, AIP. The full interview is available at https://www.aip.org/history-programs/niels-bohr-library/oral-histories.

Causality and Chance expressed the long philosophical maturing process Bohm had been experiencing since leaving Princeton at the end of 1951. The most influential reading was his ongoing study of *Science of Logic*, the early 18th century work of the German philosopher G. W. F. Hegel. Mario Schönberg was the physicist who encouraged Bohm to study Hegel, but Bohm also benefitted from discussions with Mario Bunge and Walter Schützer's brother, who was a philosophy teacher, in Argentina and Brazil, from Mashulan Groll, a left wing activist of the kibbutz movement in Israel, from Erich Burhop, and from Norman Franklin, director of the publisher Routledge and Kegan Paul.[71] The singularity of Bohm's views in this book, given his defense of the causal interpretation, was that causal laws and statistical laws appear to have the same epistemological status. Thus Bohm was weakening the prominence he had previously attributed to causal laws in science. The other idea present in the book is the infinity of levels of reality. Bohm expressed this idea in his letters as early as November 1951 and his recollections suggest he may have been inspired by the idea of the inexhaustibility of the electron, which he picked up from Lenin's works while still at Princeton. This idea had acquired a concrete physical form for him since him and Vigier had modeled the hidden-variables interpretation based on the existence of a sub-quantum level of reality which had random movements.[72] Bohm's independent thinking is evident when we note that these new ideas were not initially shared by all his fellow Marxist physicists. Évry Schatzman, speaking on behalf of Vigier and himself, wrote to Bohm "We may be wrong, but we do not agree at all with your ideas about the different levels of reality. It seems to us that it is a formal interpretation of the famous sentence of Lenin, in Materialism and Empiriocriticism, about the different levels of reality".[73]

The book was well received, as acknowledged by Bohm himself, despite the reception among philosophers not being as high as Bohm had expected. Very recently, the philosopher Boris Kožnjak surveyed the reception of *Causality and Chance* through contemporary reviews, and concluded that except for a vitriolic review by Rosenfeld, "by the end of 1957 and at the beginning of 1958 positive reviews of Bohm's book started to flourish in science and philosophy journals".[74] However, the readers did not notice that Bohm was indeed weakening the status of causal laws in science he had earlier attributed; only later commentators and Bohm himself would emphasize this. For Bohm the publication of this book meant a laissez-passer to the philosophical stage. The book created the opportunity of conversation with philosophers of science who were deeply interested in the philosophy of 20th century physics. This was mainly the cases of Paul Feyerabend, who was a visiting fellow at the Bristol Philosophy Department at the time; Karl Popper, who was in London but with whom Bohm could discuss philosophical issues of quantum mechanics;

---

[71] Interview of David Bohm by Maurice Wilkins, idem.
[72] Idem. Bohm to Miriam Yevick, 3 Nov [1951], in Talbot (2017, 204–206 l [Letter 58]). Bohm and Vigier (1954).
[73] Schatzaman to Bohm, 20 Feb 1952 (Besson 2018, p. 313).
[74] Kožnjak (2018, 93).

and Stephen Körner, chair of the Philosophy Department.[75] Körner would be the person behind the organization of the most important scientific gathering Bohm had attended in the 1950s to discuss his interpretation of quantum physics.

The Colston Society has a long history anteceding the setting up of the University of Bristol, in 1909. In fact it was part of the movement which led to the founding of the university. After World War II the then called "Colston Research Society" began to organize a series of annual symposia dedicated to different scholarly subjects, with their proceedings, including the debates, immediately published. For the 1957 symposium, the Society invited Stephen Körner to organize it dealing with the subject of philosophy of physics. Helped by Maurice Pryce, Körner adopted the subject "Observation and Interpretation—A Symposium of Philosophers and Physicists". Indeed most of the topics and speakers dealt with issues concerning the foundations of quantum mechanics. Bohm was invited to share the second session with nobody less than Léon Rosenfeld. Bohm spoke about "A proposed explanation of quantum theory in terms of hidden variables at a sub-quantum-mechanical level," along the same lines as his 1954 paper with Vigier, while Rosenfeld chose "Misunderstandings about the foundations of quantum theory". Other invitees included Jean-Pierre Vigier, who was Bohm's companion in the causal campaign, and Markus Fierz, closely connected to Wolfgang Pauli. In addition, among the speakers who directly approached issues of either interpretation or foundations of quantum mechanics, were Fritz Bopp, Hilbrand J. Groenewold, Georg Süssman, and Paul K. Feyerabend. Karl Popper was scheduled to talk but could not attend. Instead his work was read and discussed. Either from the list of participants or the titles of the talks, it was expected to be a major battle in the campaign for the causal interpretation. It was the kind of battle whose results could decide the fate of the campaign.

As the number of voices criticizing or doubting the causal interpretation soon after Bohm's papers were published were impressive, one might expect Bohm and Vigier to have been isolated at the meeting. However, as a fine analysis conducted by Boris Kožnjak has shown, almost the opposite happened. Or, at least, there was kind of a draw, without victors, which should be considered a great victory for the causal interpretation supporters. Kožnjak's analysis may be encapsulated in this fragment of his paper:[76]

> Not only that Rosenfeld did not receive wide support for his criticism of the causal quantum theory program among the speakers and participants of the Symposium, but it was he who was in fact challenged, and not only by the very dissents Bohm and Vigier, and other apostates from the Copenhagen school, like Feyerabend or Fritz Bopp, but also by the very 'orthodoxists' like Pryce, and especially by Fierz, a former Wolfgang Pauli's assistant, close friend and a colleague, not only a conservative and careful physicist as he was, but also a relentless critic of 'alternative interpretations' of quantum mechanics, who was, as we have seen, meant to be Pauli's voice at the Symposium.

---

[75]Paul Feyerabend (1960) praised Bohm's *Causality and Chance* as containing "an explicit refutation of the idea that complementarity, and complementarity alone, solves all the ontological and conceptual problems of microphysics". Popper, who had been interested in the foundations of quantum mechanics from the 1930s, only became an active protagonist in the quantum controversy in the early 1980s. See (Santo 2018), (Freire Jr. 2015), and (Popper and Bartley 1982).

[76]Kožnjak (2018, 92)

Kožnjak brought together the debates at the Colston's symposium as well as the reception of *Causality and Chance* and the reception of the symposium's proceedings to conclude that "by the end of 1950s, there seemed to be open a new wide space for the ideologically non-biased critical appraisal of Bohm's theory, with the initially almost unanimously negative reception of the theory being replaced by a more open-minded one," while remarking that "Bohm himself did not capitalize this victory right away, since by the end of 1950s he had temporarily abandoned his early causal program;" which is a subject we shall come back to in the next chapter. Many reasons may have influenced the outcome of the Colston symposium. Rosenfeld's dogmatic stands and his "unpreparedness to address the audience of the unconverted" played a role in alienating him even from close potential allies such as Fierz.[77]

The intellectual openness of the symposium, bringing together philosophers and physicists, might also have played a role. It should not be underestimated that Bohm also presented a view on quantum mechanics which was not yet set in stone, instead it was in the making. In fact, Bohm neither presented his original 1952 interpretation nor limited himself to defend the 1954 work in which he and Vigier had modified the original interpretation introducing the sub-quantum level. In the debates following the talks Bohm's closing statements began by declaring "in the original form of this theory, which was quite different, the effort was …" and went to criticize his original work, "this interpretation had many features that were unsatisfying; and especially the quantum potential was rather arbitrary in form". Following, he declared "hence I began to change the theory until I arrived at the model that I described today in which the quantum potential is an effect of a certain statistical motion". After some analogies with thermodynamics and kinetic theory, he declared: "The quantum potential comes out as an effective statistical term which should be added to the remaining energies to help determine the average motion of the electron. The quantum potential therefore ceases to play an essential role because everything is explained in a deeper way". The reference to the "deeper way" was germane to the ideas of infinite levels of reality Bohm had been developing. He concluded his reasoning touching on the other tough epistemological issue, the role of determinism in scientific laws, presenting his current views in a manner which recalls the title of the book he had just published: "Now I don't insist that the deeper laws will be purely determinate laws or purely statistical laws". Bohm concluded suggesting that "in terms of the infinity of nature we may rather say that every law is at one stage determinate, at another statistical; every law leaves out an infinity of factors on lower levels".[78] This was far from the emphasis on the "causal" interpretation which had dominated the debates so far about his early interpretation.

Bristol was also a place where Bohm resumed scientific work not directly related to the interpretation of quantum mechanics. These works were developed with Gideon Carmi and Yakir Aharonov, the two Israeli students who had accompanied him from Israel. With Carmi, Bohm resumed the work on collective variables dealing with separation of motion of many-body systems. However, these papers only were

---

[77]Citations from Kožnjak (2018, 94–95)

[78]Bohm's statements are in Körner (1957, 60)

published later, in 1964. The most influential and truly breakthrough work, however, was the second work with Aharonov, which led to the discovery of what we know today as the Aharonov-Bohm effect. Bohm's memories tell us that the first idea was entirely due to Aharonov. According to Bohm, "Aharonov got the idea at that time of considering a line of flux in a magnetic field, considering electron beams split in two, so half would go around one side and half around the other and then they would recombine to produce interference". Still according to Bohm, Aharonov also made the first analysis of this ideal experiment. For Bohm, "[Aharonov] showed that the interference pattern would be shifted according to the strength of the line of flux, even though the electron beam never came into the line of flux". Then they wrote the paper—"We wrote a paper on it"—which stirred up much debate because, "some people found it, most people found it very puzzling. Those who really understood quantum mechanics accepted it after a little thought, but for years people have refused to accept that" (Fig. 4.5).[79]

It should be noted that this work was not driven by concerns related to the reinterpretation of quantum mechanics, instead it dealt with the still unforeseen effects of standard quantum mechanics mathematical formalism. Aharonov and Bohm analyzed the role of electromagnetic potentials in quantum theory and noted that the vectorial part of the electromagnetic potential could lead to unexpected results. This was because both parts of this potential, the scalar and the vectorial, enter the Schrödinger equation, through the Hamiltonian of the system, at the same level from a physical point of view. Then they suggested the existence of a new physical effect which they illustrated with an electron beam being split around a region where an electromagnetic field is confined. They showed a beam may undergo a physical change even passing through a field-free region. They then argued that this was a quantum effect related to the vector potential, since in classical electromagnetism this potential is considered to be without physical meaning. In more general terms, they explained the difference between the classical and the quantum situations: "In classical mechanics, we recall that potentials cannot have such significance because the equation of motion involves only the field quantities themselves. For this reason, the potentials have been regarded as purely mathematical auxiliaries, while only the field quantities were thought to have a direct physical meaning". Then, they made the contrast: "In quantum mechanics, the essential difference is that the equations of motion of a particle are replaced by the Schrödinger equation for a wave. This Schrödinger equation is obtained from a canonical formalism, which cannot be expressed in terms of the fields alone, but which also requires the potentials".[80]

This paper led to a flood of experiments and theoretical explanations and is by far the most influential paper authored by David Bohm, amounting to more than

---

[79]Bohm and Carmi (1964). Aharonov and Bohm (1959). Interview of David Bohm by Maurice Wilkins on 1987 January 30, AIP, www.aip.org/history-programs/niels-bohr-library/oral-histories/32977-7. A short biographical note on Aharonov is Pines (2014).
[80]Aharonov and Bohm (1959, 490)

Fig. 4.5 Yakir Aharonov and Gideon Carmi were the two students Bohm met at Technion. They followed him to Bristol where they obtained their Ph.D. With Aharanov, Bohm discovered the Aharonov-Bohm effect. At the Quantum Mechanics Conference held at Xavier University, Cincinnati, Ohio, 1962. L-R Row: Eugene Wigner, Rosen, Dirac, Podolsky, Aharonov and Ferry; 2nd Row: Unidentified, Carmi, Unidentified, Brill, Unidentified and Yanase; 3rd Row: Unidentified, Unidentified, Merzbacher, Unidentified, Unidentified, Guth, Shimony, Unidentified and Unidentified. *Credits* Babst Photography, courtesy of AIP Emilio Segrè Visual Archives, Gift of Abner Shimony

4200 citations as of September 2018.[81] It brought wide recognition to both, which included the 1998 Wolf Prize for Aharonov.

At the end of his stay at Bristol with a research fellowship, Bohm was appointed to a new chair of theoretical physics at Birkbeck College, in London, in 1960. Birkbeck had been set up in 1823 as a college for evening courses for workers. It was never a top-tier traditional British university but it soon revealed its excellence through research and teaching. Its quality may be inferred from the presence of, among its faculty, historian Eric Hobsbawm, philosopher Slavoj Zizek, poet T. S. Eliot, biophysicist Rosalind Franklin, and physicists John D. Bernal and Patrick Blackett.

---

[81] Aharonov and Bohm (1959). (Peshkin and Tonomura 1989). Number of citations from the Web of Science, accessed o 05 Sep 2018. For the debate on the theoretical interpretation of the Aharonov-Bohm effect, see (Lyre 2009).

The leading British scientific magazine, *Nature*, announced Bohm's appointment along the following lines:[82]

> Prof. David Bohm, whose appointment is announced to the new chair of theoretical physics at Birkbeck College, London, is well known among physicists as the originator of the Bohm-Pines theory of collective motions in a plasma. He is known to a wider public for his views on the philosophy of science, particularly his criticisms of the interpretation of quantum theory advocated by the school of Niels Bohr.

After a short CV, which included a reference to his war work on "the electromagnetic separation of isotopes" but no mention of the McCarthyism issue, *Nature* concluded "he is a physicist of great inventiveness and imagination, whose stimulus and leadership have set several others on the pathway to success in theoretical physics". Bohm was thus being enthroned in British academia both as a mainstream physicist as well as a quantum dissident, both related to his intellectual skills and professional leadership.

References

Aharonov, Y., Bohm, D.: Significance of electromagnetic potentials in the quantum theory. Phys. Rev. **115**(3), 485–491 (1959)

Anderson, P.: Considerations on Western Marxism. NLB, London (1976)

Andrade, A.M.R.: Físicos, mésons e política: a dinâmica da ciência na sociedade. HUCITEC, São Paulo (1999)

Berry, M., Pollard, B.: The physical tourist physics in bristol. Phys. Perspect. **10**, 468–480 (2008)

Besson, V.: L'interprétation causale de la mécanique quantique: biographie d'un programme de recherche minoritaire (1951–1964), Ph.D. Dissertation, Université Claude Bernard Lyon 1 and Universidade Federal da Bahia, (2018)

Bitbol, M.: Schrödinger's Philosophy of Quantum Mechanics. Kluwer, Dordrecht (1996)

Bohm, D.: Proof that probability density approaches $|\psi|2$ in causal interpretation of the quantum theory. Phys. Rev. **89**(2), 458–466 (1953a)

Bohm, D.: A Discussion of Certain Remarks by Einstein on Born's Probability Interpretation of the ψ—Function, pp. 13–19. Scientific Papers presented to Max Born. Edinburgh, Oliver and Boyd (1953b)

Bohm, D.: Causality and chance in modern physics. Foreword by Louis De Broglie. London, Routledge and Paul (1957)

Bohm, D.: Hidden variables in the quantum theory. In: Bates, D.R. (ed.) Quantum Theory—III—Radiation and High Energy Physics, pp. 345–387. Academic, New York (1962)

Bohm, D.: Wholeness and the Implicate Order. Routledge and Kegan Paul, London (1980)

Bohm, D., Aharonov, Y.: Discussion of experimental proof for the paradox of Einstein, Rosen, and Podolsky. Phys. Rev. **108**(4), 1070–1076 (1957)

Bohm, D., Carmi, G.: Separation of motions of many-body systems into dynamically independent parts by projection onto equilibrium varieties in phase space. I. Phys. Rev. **133**(2A), A319–A331 (1964)

Bohm, D., Hiley, B.J.: The Undivided Universe: An Ontological Interpretation of Quantum Theory. Routledge, London (1993)

---

[82]*Nature*, 22 April 1961, p. 308.

Bohm, D., Hillion, P., Takabayasi, T., Vigier, J.P.: Relativistic rotators and bilocal theory. Prog. Theor. Phys. **23**(3), 496–511 (1960a)

Bohm, D., Hillion, P., Vigier, J.P.: Internal quantum states of hyperspherical (Nakano) relativistic rotators. Prog. Theor. Phys. **24**(4), 761–782 (1960b)

Bohm, D., Schiller, R.: A causal interpretation of the Pauli equation (B). Nuovo Cimento Suppl **1**(1), 67–91 (1955)

Bohm, D., Schiller, R., Tiomno, J.: A causal interpretation of the Pauli equation (A). Nuovo Cimento Suppl **1**, 48–66 (1955)

Bohm, D., Schützer, W.: The general statistical problem in physics and the theory of probability. Supplemento del Nuovo Cimento **2**(4), 1004–1047 (1955)

Bohm, D., Vigier, J.P.: Model of the causal interpretation of quantum theory in terms of a fluid with irregular fluctuations. Phys. Rev. **96**(1), 208–216 (1954)

Bohm, D., Vigier, J.P.: Relativistic hydrodynamics of rotating fluid masses. Phys. Rev. **109**(6), 1882–1891 (1958)

Bohr, A., Mottelson, B.R., Pines, D.: Possible analogy between the excitation spectra of nuclei and those of the superconducting metallic state. Phys. Rev. **110**(4), 936–938 (1958)

de Broglie, L.: La théorie de la mesure en mécanique ondulatoire interprétation usuelle et interprétation causale. Gauthier-Villars, Paris (1957)

de Broglie, L., Vigier, J.P., Bohm, D., Takabayasi, T.: Rotator model of elementary particles considered as relativistic extended structures in Minkowski space. Phys. Rev. **129**(1), 438–450 (1963)

Brownell, G.L.: Physics in South America. Physics today, pp. 5–12 (1952)

Bunge, M.: What is chance? Sci. Soc. **15**, 209–231 (1951)

Carson, C.: Heisenberg in the Atomic Age: Science and the Public Sphere. German Historical Institute, Washington, DC (2010)

Caute, D.: Le Communisme et les intellectuels français, 1914–1966. Gallimard, Paris (1967)

Cross, A.: The crisis in physics: dialectical materialism and quantum theory. Soc. Stud. Sci. **21**, 735–759 (1991)

Duarte, F.J.: The origin of quantum entanglement experiments based on polarization measurements. Eur. Phys. J. H **37**(2), 311–318 (2012)

Einstein, A.: Elementare Überlegungen zur Interpretation der Grundlagen der Quanten-Mechanik. Scientific Papers presented to Max Born, pp. 33–40. Oliver and Boyd, Edinburgh. French translation in Albert Einstein, Oeuvres Choisies, ed. by F. Balibar, B. Jech, and O. Darrigol, Paris: Editions du Seuil-CNRS, 1989, pp. 1251–1256 (1953)

Einstein, A., Born, M., Born, H.: The Born-Einstein letters: correspondence between Albert Einstein and Max and Hedwig Born from 1916–1955, with commentaries by Max Born. Macmillan, London (1971)

Epstein, S.T.: The causal interpretation of quantum mechanics. Phys. Rev. **89**(1), 319 (1953a)

Epstein, S.T.: The causal interpretation of quantum mechanics. Phys. Rev. **91**(4), 985 (1953b)

Feyerabend, P.: Professor Bohm's philosophy of nature. Br. J. Philos. Sci. **10**(40), 321–338 (1960)

Feynman, R.: The present situation in fundamental theoretical physics. An. Acad. Bras. Cienc. **26**(1), 51–60 (1954)

Fock, V.A.: On the interpretation of quantum mechanics. Czechoslov. J. Phys. **7**, 643–656 (1957)

Forstner, C.: The early history of David Bohm's quantum mechanics through the perspective of Ludwik Fleck's thought-collectives. Minerva **46**(2), 215–229 (2008)

Freire Junior, O.: The crisis in physics—comment. Soc. Stud. Sci. **22**(4), 739–742 (1992)

Freire Junior, O.: David Bohm e a controvérsia dos quanta. Centro de Lógica, Epistemologia e História da Ciência, Campinas, Brazil (1999)

Freire Junior, O.: Science and exile: David Bohm, the cold war, and a new interpretation of quantum mechanics. Hist. Stud. Phys. Biol. Sci. **36**(1), 1–34 (2005)

Freire Junior, O.: The Quantum Dissidents—Rebuilding the Foundations of Quantum Mechanics 1950–1990. Springer, Berlin (2015)

Freire Jr., O.: On the connections between the dialectical materialism and the controversy on the quanta. YearB. Eur. Cult. Sci. **6**, 195–210 (2011)

Freire Junior, O., Lehner, C.: 'Dialectical materialism and modern physics', an unpublished text by Max Born. Notes Rec. R. Soc. **64**(2), 155–162 (2010)

Freistadt, H.: The crisis in physics. Sci. Soc. **17**, 211–237 (1953)

Freistadt, H.: Connection between recent theories of Bohm, de Broglie, Dirac, and Schrodinger. Phys. Rev. **98**(4), 1176 (1955)

Freistadt, H.: The causal formulation of quantum mechanics of particles: the theory of de Broglie. Bohm and Takabayasi. Nuovo Cimento Suppl **5**, 1–70 (1957)

Frost, R.: The road not taken; an introduction to Robert Frost. A selection of Robert Frost's poems with a biographical pref. and running commentary by Louis Untermeyer. Holt, New York (1951)

Gaddis, J.L.: The Cold War: A New History. Penguin, New York (2005)

Graham, L.R.: Science, Philosophy, and Human Behavior in the Soviet Union. Columbia University Press, New York (1987)

Heisenberg, W.: Physics and Philosophy; the Revolution in Modern Science. Harper, New York (1958)

Halpern, O.: A proposed re-interpretation of quantum mechanics. Phys. Rev. **87**(2), 389 (1952)

Hobsbawm, E.J.: How to Change the World: Marx and Marxism, 1840-2011. Little, Brown, London (2011)

Hoffmann, D.: Robert Havemann: antifascist, communist, dissident. In: Macrakis, K., Hoffmann, D. (eds.) Science Under Socialism: East Germany in Comparative Perspective, pp. 269–285. Harvard University Press, Cambridge, MA (1999)

Home, D., Whitaker, M.A.B.: Ensemble interpretations of quantum-mechanics—A modern perspective. Phys. Rep. **210**(4), 223–317 (1992)

Isserman, M.: The 1956 Generation—An alternative approach to the history of American communism. Radical Am. **14**(2), 43–51 (1980)

Jacobsen, A.: Léon Rosenfeld's Marxist defense of complementarity. Hist. Stud. Phys. Biol. Sci. **37**(Suppl), 3–34 (2007)

Jacobsen, A.: Léon Rosenfeld—Physics, Philosophy, and Politics in the 20th Century. World Scientific, Singapore (2012)

Kaiser, D.: How the hippies saved physics: science, counterculture, and the quantum revival. W. W Norton, New York (2012)

Keller, J.B.: Bohm's interpretation of the quantum theory in terms of Hidden variables. Phys. Rev. **89**(5), 1040–1041 (1953)

Kojevnikov, A.: David Bohm and collective movement. Hist. Stud. Phys. Biol. Sci. **33**, 161–192 (2002)

Kojevnikov, A.: Stalin's Great Science: The Times and Adventures of Soviet Physicists. Imperial College Press, London (2004)

Kojevnikov, A.: Probability, marxism, and quantum ensembles. Yearb. Eur. Cult. Sci. **6**, 211–235 (2011)

Körner, S. (ed).: Observation and Interpretation; A Symposium of Philosophers and Physicists. Butterworths, London (1957)

Kožnjak, B.: The missing history of Bohm's hidden variables theory: the ninth symposium of the Colston research society, Bristol, 1957. Stud. Hist. Philos. Mod. Physics **62**, 85–97 (2018)

Kuzemsky, A.L.: Works by D. I. Blokhintsev and the development of quantum physics. Phys. Part. Nucl. **39**(2), 137–172 (2008)

Lopes, J.L.: Richard Feynman in Brazil: personal recollections. Quipu **7**, 383–397 (1990)

Lyre, H.: Aharonov-Bohm effect. In: Greenberger, D., Hentschel, K., Weinert, F. (eds.) Compendium of Quantum Physics: Concepts, Experiments, History and Philosophy, pp. 1–3. Springer, Berlin (2009)

Maia Filho, A.M., Silva, I.: O experimento WS de 1950 e as suas implicações para a segunda revolução da mecânica quântica. Revista Brasileira de Ensino de Física **41**(2) (2019)

Martinez, J-P.: Vladimir Fock (1898–1974): itinéraire externaliste d'une pensée internaliste—Antiréductionnisme et réalisme scientifique en physique modern, Ph.D. Dissertation, Université Sorbonne Paris Cité & Université Paris Diderot, (2017)

Mehra, J.: The Beat of a Different Drum: The Life and Science of Richard Feynman. Clarendon, Oxford (1994)

Moore, W.J.: Schrödinger: Life and Thought. Cambridge University Press, Cambridge, England (1989)

New research techniques in physics.: Rio de Janeiro and São Paulo, July 15–29, 1952 (1954)

Ory, P., Sirinelli, J.-F.: Les Intellectuels en France de l'affaire Dreyfus a' nos jours. Perrin, Paris (2004)

Paty, M.: Sur les 'variables cachées' de la mécanique quantique—Albert Einstein, David Bohm et Louis de Broglie. La pensée **292**, 93–116 (1993)

Paty, M.: The nature of Einsteins objections to the Copenhagen interpretation of quantum mechanics. Found. Phys. **25**(1), 183–204 (1995)

Pauli, W.: Remarques sur le problème des paramètres cachés dans la mécanique quantique et sur la théorie de l'onde pilote. In: George, A. (ed.) Louis de Broglie—physicien et penseur, pp. 33–42. Editions Albin Michel, Paris (1953a)

Pauli, W., Meyenn, K.V.: Wissenschaftlicher Briefwechsel mit Bohr, Einstein, Heisenberg u. a. Band IV Teil I 1950–1952. Springer, Berlin (1996)

Pauli, W., Meyenn, K.V.: Wissenschaftlicher Briefwechsel mit Bohr, Einstein, Heisenberg u. a. Band IV Teil II 1953–1954. Springer, Berlin (1999)

Peat, F.D.: Infinite Potential: The Life and Times of David Bohm. Addison Wesley, Reading, MA (1997)

Pechenkin, A.: The early statistical interpretations of quantum mechanics in the USA and USSR. Stud. Hist. Philos. Mod. Phys. **43**(1), 25–34 (2012)

Pechenkin, A.: Leonid Isaakovich Mandelstam: Research, Teaching. Life. Springer, New York (2013)

Peshkin, M., Tonomura, A.: The Aharonov-Bohm Effect. Springer, Berlin (1989)

Pinault, M.: Frédéric Joliot-Curie. O. Jacob, Paris (2000)

Pines, A., Aharonov, Y.: Thinking quantum, In: Struppa, D.C., Tollaksen, J.M. (eds.) Quantum Theory: A Two-Time Success Story—Yakir Aharonov Festschrift, pp. 399–405, Springer, New York (2014)

Popper, K.R., Bartley, W.W.: Quantum Theory and the Schism in Physics. Rowan and Littlefield, Totowa, NJ (1982)

Rattner, H.: Tradição e mudança (a comunidade judaica em São Paulo). Atica, São Paulo (1977)

Rodrigues, L.M.: O PCB: Os dirigentes e a organização. In B. Fausto (org.) História geral da civilização brasileira, tomo III, vol. 3: O Brasil republicano—sociedade e política (1930–1964), pp. 361–443. Bertrand Brasil, Rio de Janeiro: 361–443 (1996)

Rosenfeld, L.: Physics and metaphysics. Nature **181**(4610), 658–658 (1958)

Rosenfeld, L.: L'évidence de la complementarité, in A. George (ed.), Louis de Broglie—physicien et penseur, pp. 43–65. Editions Albin Michel, Paris pp. 43–65 (1953). [A slightly modified English version of this paper is Strife about complementarity, Science progress. 163, 1393–1410 (1953), reprinted in Robert Cohen and John Stachel (eds.). Selected papers of Léon Rosenfeld (Dordrecht, D. Reidel, 1979)

Rosenfeld, L.: A filosofia da física atômica. Ciência e cultura **6**(2), 67–72 (1954)

Rosenfeld, L.: Heisenberg, physics and philosophy. Nature **186**, 830–831 (1960)

Rosenfeld, L.: Berkeley redivivus. Nature **228**, 479 (1970)

Rosenfeld, L.: Classical Statistical Mechanics. Livraria da Física & CBPF, São Paulo (2005)

Saidel, R.G., Plonski, G.A.: Shaping modern science and technology in Brazil. The contribution of refugees from national socialism after 1933. Leo Baeck Inst. YearB **1994**, 257–270 (1994)

Santo, F. del.: Genesis of Karl Popper's EPR-like experiment and its resonance amongst the physics community in the 1980s. Studies in History and Philosophy of Modern Physics 62, 56–70 (2018)

Schatzman, E.: Physique quantique et realité. La pensée **42–43**, 107–122 (1953)

Schönberg, M.: On the hydrodynamical model of the quantum mechanics. Il Nuovo Cimento, XII **1**, 103–133 (1954)

Schönberg, M.: Quantum theory and geometry. In: Kockel, B., Macke, W., Papapetrou, A. (eds.) Max-Planck-Festschrift 1958, pp. 321–338. Deutscher Verlag der Wissenschaften, Berlin (1959) [Reprinted in M. Schönberg, Obra Científica de Mario Schönberg, São Paulo: EDUSP, 2013

Schönberg, M., Hamburger, A.I.: Obra Científica de Mario Schönberg, vol. 1, pp. 1936–1948. EDUSP, São Paulo (2009)

Schönberg, M., Hamburger, A.I.: Obra Científica de Mario Schönberg, vol. 2, pp. 1949–1987. EDUSP, São Paulo (2013)

Schrecker, E.: The age of McCarthyism—A brief history with documents. Bedford/St. Martins's, Boston (2002)

Schrödinger, E.: Are there quantum jumps? I and II. Br. J. Philos. Sci. **3**, 109–123 (1952). 233–242

Schrödinger, E.: The meaning of wave mechanics. In: George, A. (ed.) Louis de Broglie—physicien et penseur, pp. 16–32. Albin Michel, Paris (1953)

Schrödinger, E., Bitbol, M.: Physique quantique et représentation du monde introd. et notes par Michel Bitbol. Ed. du Seuil, Paris (1992)

Schweber, S.S.: QED and the Men Who Made It: Dyson, Feynman, Schwinger, and Tomonaga. Princeton University Press, Princeton, NJ (1994)

Shannon, D.A.: The Decline of American Communism—A History of the Communist Party of the United States since 1945. The Chatham Bookseller, Chatham, NJ (1959)

Stöltzner, M.: What John von Neumann thought of the Bohm interpretation. In: Greenberger, D., Reiter, W.L., Zeilinger, A. (eds.) Epistemological and Experimental Perspectives on Quantum Mechanics, pp. 257–262. Springer, Dordrecht (1999)

Takabayasi, T.: On the formulation of quantum mechanics associated with classical pictures. Prog. Theor. Phys. **8**(2), 143–182 (1952)

Takabayasi, T.: Remarks on the formulation of quantum mechanics with classical pictures and on relations between linear scalar fields and hydrodynamical fields. Prog. Theor. Phys. **9**(3), 187–222 (1953)

Talbot, C. (ed).: David Bohm: Causality and Chance, Letters to Three Women, Springer, Berlin (2017)

Vals, A.V.: Louis de Broglie et la diffusion de la mécanique quantique en France (1925–1960), Ph.D. dissertation, Université Claude Bernard Lyon 1 (2012)

Vieira, C.L., Videira, A.A.P.: Carried by history: Cesar lattes, nuclear emulsions, and the discovery of the Pi-meson. Phys. Perspect. **16**(1), 3–36 (2014)

Von Neumann, J.: Mathematical Foundations of Quantum Mechanics. Princeton University Press, Princeton, NJ (1955)

Wang, J.: American Science in an Age of Anxiety: Scientists, Anticommunism, and the Cold War. University of North Carolina Press, Chapel Hill, NC (1999)

Westad, O.A.: The Cold War—a World History. Basic Books, New York (2017)

Wu, C.S., Shaknov, I.: The angular correlation of scattered annihilation radiation. Phys. Rev. **77**(1), 136 (1950)

Whyte, L.L.: The scope of quantum-mechanics—discussion. Br. J. Philos. Sci. **9**(34), 133–134 (1958)

Chapter 5
From the Causal Interpretation to the Wholeness and Order—The First Stage of the London Years (1960–1979)

In hindsight, the thirty years David Bohm spent in London, as a Professor at Birkbeck College, were the calmest years of his life, at least compared to the anxieties of the previous decades. However, the first stage of his years in London was marked by meaningful intellectual changes, both in science and philosophy. He dropped his campaign for the causal interpretation and looked for a different epistemology and worldview in order to understand the quantum. It was not a quick transition as only in early 1970s the ideas of wholeness and order coalesced as a research program. While he kept up his duties as physics professor, his attraction to physics as it was being practiced dwindled, in particular due to his estrangement with plasma and solid physics as they were being practiced. Last but not least, he moved from a Marxist commitment with an emphasis on social and economic determinants to the appreciation of spiritual aspects and the world of ideas, which included readings and relationships with mystics and with the Indian writer Jiddu Krishnamurti. Considering all these changes together, they meant David Bohm was experiencing not only an intellectual shift but also an existential crisis. Still, in early 1960s Bohm realized the full consequences of citizenship other than American as the US government withdrew his. In the midst of this, around 1970, Bohm followed with delight a surge of interest—among the younger physicists—in the debate on the foundations of quantum mechanics. Most of this interest was driven by the appearance of a theorem, formulated by John Bell, contrasting local hidden variables to plain quantum mechanics. The theorem was followed by experiments to settle the contrast. This stage was closed in the late 1970s with a ruse of history. The hidden variables approach and its quantum potential, which he had abandoned since the early 1960s, resurfaced in the hands of some of his students. This twist would lead him to a new stage in his long quest to understand the quantum, the new stage being that of the attempts to accommodate the quantum potential into the wholeness research program.

In this chapter thematic as well as chronological order are mixed. Section 5.1 is dedicated to the existential crisis Bohm underwent in early 1960s, which meant the abandonment of the causal program and priority for causality as well as an abatement

© Springer Nature Switzerland AG 2019
O. Freire Junior, *David Bohm*, Springer Biographies,
https://doi.org/10.1007/978-3-030-22715-9_5

of interest in physics. Section 5.2 deals with Bohm's rapprochement with mystics and the Indian writer Jiddu Krishnamurti. The relationship with Krishnamurti would not abate for more than two decades. In Sect. 5.3 we present Bohm's unsuccessful attempts to regain his US citizenship. Section 5.4 deals with a wider subject, Bohm's musings in physics and epistemology looking for ways to approach the quantum phenomena. In this section we feature his work with Jeffrey Bub, who was his doctoral student and continued working with Bohm for a while, and the interactions with Rosenfeld and Donald Schumacher. Further, we comment on the textbook on relativity and its reception. The fifth section is devoted the coalescence of the ideas of implicate order and wholeness as the pillars of Bohm's new approach to the quanta. The final section, the sixth, temporally overlaps with the previous ones as it presents the intellectual and professional changes among physicists concerning the research on the foundations of quantum mechanics, most of which occurred around 1970.

5.1   Abandonment of the Causal Program, Abandonment of Priority for Causality, and Abatement of Interest in Physics

Since early 1957 Bohm had signaled how aware he was of the deficiencies of his suggested 1952 reinterpretation of quantum mechanics. In the paper with Aharonov, Bohm and Aharonov (1957), where they checked the EPR predictions against the available data from the Wu and Sakhnov experiment, he declared "it must be admitted, however, that this quantum potential seems rather artificial in form." In addition, he noted that this potential was "subject to the criticism [...] that it implies instantaneous interactions between distant particles, so that it is not consistent with the theory of relativity." Instead of supporting his causal interpretation Bohm suggested a new interpretation, based on the assumption of a "deeper subquantum-mechanical level," an idea he referred to in his book *Causality and Chance*, published the same year. At the Bristol conference we commented on in the previous chapter, his choice of talk was the same general idea of the subquantum level and not the defense of the 1952 approach. As a matter of fact, the idea of an infinity of levels and laws remained a worldview but a detailed presentation of the quantum sublevel, in terms of a conceptual model or a physical treatment, never came out. The closest Bohm had arrived was in the 1954 paper with Jean-Pierre Vigier where they reinterpreted quantum mechanics through a hydrodynamical model with the introduction of particles randomly moving at a subquantum level. In the early 1960s Bohm received an invitation to write a chapter on hidden variables in quantum theory for the 3-volume collection D. Bates was editing for an overview of the state of the art on the development of quantum theory. It was a nice opportunity to relaunch the campaign for the causal interpretation, had Bohm remained committed to this approach. Indeed, this opportunity was seen as such on the opposite side of the battle concerning the interpretation of quantum mechanics. Historian Anja Jacobsen narrated that physicists in Copenhagen, particularly Léon Rosenfeld and Stefan Rozental, were very concerned with the inclusion of this chapter in a collection dedicated to a wider audience. According to Jacobsen, "Rosenfeld burst out to the editor, D. R. Bates," on May 22, 1957, in the following terms: "Could this possibly mean that you are smitten with Bohmitis? [...] Bohm's ideas apart from being demonstrably wrong in many points, are entirely unscientific in as much as they are based on metaphysical speculation, not supported by any evidence whatsoever."[1]

Bohm's text however, was less dramatic than Rosenfeld could have expected. He presented a very pedagogical review concerning the introduction of hidden variables, including his own approach, and then systematically reviewed the criticisms such an approach had received. Thus the notion of quantum potential was not "entirely satisfactory" because it was strange and arbitrary as it has no sources. While this criticism is only concerned with the plausibility of the notion, "we evidently cannot be satisfied with accepting such a potential in a definite theory." Plausibility was also a problem, according to Bohm, when many-body systems were considered as there are no physical explanation for the instantaneous interactions among them which are introduced by the quantum potential. Bohm's line of defense was that the very creation of interpretations as his own interpretation had played a positive role in physics. As his interpretation was logically consistent while physically implausible, it had exhibited the faults in von Neumann's proof and the received view against the very possibility of introducing such hidden variables. Finally, Bohm conjectured, as in his 1952 paper, that in domains such as high energies and small distances, where quantum field theories present difficulties to their infinities, a modified version of a hidden variables interpretation may be useful. Bohm concluded his analysis stating that "from the discussion given in the previous section, it is clear that our central task is to develop a new theory of hidden variables." In the final sections of this paper however, Bohm presented only general considerations about how to make compatible such an approach with certain quantum features.[2]

In later recollections Bohm used the same justification for abandoning the causal interpretation; thus, in 1987, for this 70th anniversary Festschrift, he wrote: "because I did not see clearly, at the time, how to proceed further, my interests began to turn in other directions." As important as such statements are the fact is that throughout the 1960s and mid 1970s Bohm no longer published on his early proposal of reinterpreting quantum mechanics in terms of hidden variables. As we will see, even when the term "hidden variables" appeared, as in his papers with Jeffrey Bub or his papers on Bell's theorem, their contents were unrelated to his 1950s reinterpretation.[3]

There was another feature of the causal interpretation Bohm was also abandoning, this concerns the priority given to causal laws over statistical laws. While this was a philosophical feature, related to the type of scientific theories we should look for, and not a specific feature of his hidden variable physical model, relativizing causality weakened his beliefs in this physical approach as they were intertwined. Bohm had

---

[1]Bohm (1957) and Jacobsen (2012, 306).

[2]Bohm (1962, 359–363).

[3]Bohm (1987, 40).

been moving in this direction since 1953, particularly motivated by his readings of Hegel and the influence of the Brazilian physicist Mario Schönberg and the Israeli philosopher and activist Mashulan Groll. Bohm's mature view on the role of both causal and statistical laws in scientific theories appeared in print in his 1957 *Causality and Chance* book. However, the change in status of the causal laws, compared with his 1952 causal interpretation, was not emphasized by the book's reviewers. It was a reader, the American artist Charles Biederman, who noted this shift, as he preferred the previous priority for causal laws, in the midst of an extensive correspondence he and Bohm maintained for years. The letters—over 4000 pages between March 1960 and April 1969—slept for years among Bohm's correspondence till the Finish philosopher Paavo Pylkkänen began to edit them and noticed the importance of this exchange concerning the role of causality in science. Thus, it is only in hindsight that we may present Biederman as the town crier of Bohm's shift of views about the centrality of the role of causality in physical theories. As it happened, after the publication of *Causality and Chance*, Bohm deepened the relativization of the causal laws in general, and determinism in particular, and this further relativization was not independent of his break with Marxism, which was the major ideological change he had experienced in the late 1950s, after the publication of the book.

Bohm's intellectual turn, as it appeared in these letters, was acutely noted by Pylkkänen: "Here we have Bohm, who is internationally known as a defender of a deterministic interpretation of the quantum theory, and thus for many a defender of strict determinism in nature, arguing strongly for the objective existence of properties such as contingency, chance, determinism, etc. Of course, Bohm does this already in *Causality and Chance*, but here the point is made more vividly, given that Bohm is defending the role of indeterminism rather than questioning it, as he most famously did in his 1952 papers."[4] Bohm had been approached by Biederman after his reading of *Causality and Chance*. In his very first letter, in 1960, Biederman was clear-cut in his defense of determinism: "To explain my interest in your book [*Causality and Chance*]. To put it briefly, the notion of indeterminism has always seemed contrary to experience, which, even after reading your very fine book, I cannot accept even as an eventually limiting case." And yet, "I sympathize with your belief that a deeper penetration will reveal a nature of causality. But there is the possibility that this will also dispel the basis for the present 'lawless' view of nature and, rather than make it a limited case, will dispense with it entirely." Bohm's first answer to Biederman was based on a conceptual reflection about time. "Thus, there is some ambiguity in past and future. We experience this ambiguity in certain ways directly. For when we try to say 'now,' we find that by the time we have said it, the time that we meant is already past, and no longer 'now.'" He continues, citing an example closer to physics, "and if we try to do it with clocks, so as to be more precise, quantum theory implies that a similar ambiguity would arise because of the quantal structure of matter. In fact, there is no known way to make an unambiguous distinction between past and future." Thus, "it becomes impossible that the past shall completely determine the future, if only because there is no way to say unambiguously what the past really was until

---

[4]Paavo Pylkkänen's statement is in the introduction of Bohm et al. (1999, p. xix).

we know its future." As Biederman might have compared that letter with the book which was the catalyst of their correspondence, Bohm anticipated this, "as you may perhaps have noticed, my ideas on determinism and indeterminism have developed since I wrote *Causality and Chance*, although what I now think about these questions was, to a considerable extent, implicit in the point of view expressed in the book." His conclusion, in short, is that "neither determinism nor indeterminism (causality or chance) is absolute. Rather, each is just the opposite side of the whole picture," and that "in the question of determinism versus indeterminism, there is as I have said, a necessary complementary relation of the two ideas."[5]

As the exchange continued and the role of determinism in society entered the conversation, Bohm went to criticize both Marxism and its strong role in determinism. Curiously, after Bohm's break with Communism he made few references to Marxist ideas, even privately. Thus, the letters exchanged with Biederman are a unique source to identify Bohm's changes of worldviews. According to him, writing to Biederman, "for they [Marxists] felt that by studying the evolutionary process of the past, they could pick out the main direction in which history was moving. They became so attached to their theories that they were unable to review their own role objectively, or to admit new and unexpected developments not fitting into these theories." How much Marx's historical materialism depends on adopting determinism in history is debatable, however. For the purposes of our analysis, nonetheless, it is enough to consider that Bohm's rupture with Marxism may have destroyed his general belief in determinism as a feature of society and its history.[6] Indeed, the connection between the break with Marxism and abandonment of determinism in science, particularly in physics, and not only in society, in Bohm's thoughts is just a guess, albeit a plausible one.

If we take together Bohm's move from the hidden variables interpretation and the centrality of causal laws in physics, it comes as no surprise that his alliance with Jean-Pierre Vigier and Louis de Broglie, cemented as it was on these grounds, would be breached. Indeed, from the late 1950s on, the collaboration was frozen, slowly becoming a kind of a friendly divorce. Only the physicists deeply engaged with the research on the interpretations of quantum mechanics would realize what was happening among the old allies.

In parallel to his abandonment of the causal interpretation and the centrality for causality in physical theories, Bohm also had lost interest in physics as it was being practiced. This was mainly related to the work on plasma with the collective variables approach, the subject Bohm had worked on in the late 1940s with his two students, Gross and Pines, and had obtained results which were considered breakthroughs.

---

[5]Biederman to Bohm, 6 March 1960; Bohm to Biederman, 24 April 1960; both in (Bohm et al. 1999, pp. 3–4 and 8–19).

[6]Bohm to Biederman, 2 February 1961, (Bohm et al. 1999, p. 95). As the historian Eric Hobsbawm remarked, at least two features of Marxism should not be abandoned unless one gives up historical materialism as a way to change the world: (a) the triumph of socialism is the logical end of all historical evolution until the present, and (b) socialism marks the end of prehistory as it cannot and will not be an antagonistic society (Hobsbawm 1997, Chap. 11).

These achievements were due not only to the advances in the study of plasma, with Gross, but also to their extensions to the study of metals, with Pines, as well as to superconductivity and nuclear physics, the first resulting from works by Pines with John Bardeen and the second by Pines and Ben Mottelson and Aage Bohr. However, Bohm had left aside this subject when he moved to Brazil, not only because his energies were fully dedicated to the causal interpretation and the philosophical studies, but also because he became disinterested in the subject. This was either because of its applications or because of the contrast some physicists used to make between the two subjects: collective variables, which was well accepted, and reinterpretation of quantum mechanics, of which the same people were cautious. In addition, Bohm was most refractory to application concerning nuclear physics, surely due to its obvious connections with the whole affair with the HUAC in the US. The best evidence we have of Bohm's disinterest is that he refused an offer of a fellowship from the British physicist Nevill F. Mott because it was to work on the collective variables approach. Bohm refused this offer not because it was a fellowship and not a position but because he did not enjoy the connection of the offer with the subject to be worked on. And yet, this happened at the moment he had decided to leave the US and was moving to Brazil in 1951. His attitude towards the subject is well documented in this letter to Melba Phillips[7]:

> Last year, I got a fellowship offer of £600 from Mott, provided that I was willing to work on the collective theory of electron gases. I suspect that he knew about my work on the quantum theory and didn't want any work on such notions going on in his university. This sort of thing makes me a bit angry. Various people, when I ask them about quantum theory, say they are interested in this collective theory (in a rather pointed way), saying that it has promise of various applications, never failing to mention nuclear physics. Every time I hear this sort of thing, I cannot avoid a reaction of loss of interest in this collective theory, because every time I hear the word "nuclear physics", it calls up to my mind an image of the most boring possible subject in the world. The surest way to discourage me from working on the quantum theory would be to continually remind me that it might be useful in nuclear physics.

However, at the end of the 1950s Bohm's conflicting views about the collective variables approach waned and he returned to it. He lectured at the 1958 Les Houches summer school on the "General Theory of Collective Coordinates," and gave talks the same year at the University of Rome, on plasma physics, and at Bristol, on "General Theory of Collective Coordinates." This was also the subject of his work with Gideon Carmi, the second Israeli student who followed him to Bristol, the first being Yakir Aharonov. While the joint papers by Bohm and Carmi were only published in 1964, Bohm had preliminary results and took them to a conference on plasma held in Utrecht in 1959. Unfortunately, so far, we have been unable to recover any correspondence or other material from this episode. However, Bohm reported the episodes in the Utrecht conference in two different interviews, in 1981 and 1987, with Lillian Hoddeson and Maurice Wilkins, and the content of the two recollections

---

[7]Bohm to Melba Phillips, [end of 1951], in Talbot (2017, 142–143).

are pretty similar. In addition, Sarah Woolfson, Bohm's wife, who went with him to Holland, recalled attending one of the events reported by Bohm, which was recorded in the interview with Hoddeson.[8]

To Lillian Hoddeson, in 1981, he recalled: "I remember getting to a conference in Holland on this plasma work, and I felt that there was no physics there at all, they were just putting formulae on the board. They were not really interested in questions of what is the collective and what is the individual and things like that, you see. And I remember having an argument with a fellow called [Braut], and I said, 'I don't understand what's going on here,' and he said, 'I don't understand what you mean by not understanding.'" The anecdote on "understanding" was testified by Sarah Woolfson. Later on, to Maurice Wilkins, in 1987, Bohm was more specific. The formulae which were at stake in the conference in Holland used Feynman's diagrams instead of Schwinger's canonical transformation approach, which Bohm had used earlier. Bohm was also disappointed with the poor reception the work with Carmi was obtaining: "Then, later when I saw this mathematics which Carmi had done on plasma, that made no impression, I began to gradually lose interest in the subject, because I said, "They do not want to consider any ideas at all, not even mathematical ideas." More than disappointed, Bohm felt out of place at the Utrecht physics conference[9]:

> When I got there, I found it was strange, you see. First of all, I listened to the papers and there was nothing, they just filled the board with equations and I found it boring, because they did not even calculate any experiments. They just talked about their equations. It seemed that they had changed over from the method of what I called canonical transformations, which I had been working on, too. They were now called diagrams, Feynman diagrams, which are sort of a way of working out certain calculations quickly. Somehow physicists felt much happier about that. They felt confident that they were talking about something with these diagrams, that they understood it. But I felt that they did not, that these were just technical devices for manipulating the formulae.

Bohm's recollections are punctuated with signs of distance from his fellow physicists. The conference was attended by Soviet physicists, led by the eminent Nikolay N. Bogolyubov, which was as a sign of the détente between the USSR and the US arriving to western and eastern scientists in the context of the Cold War. Bohm's recollections were of no interaction with those physicists. As he reported to Wilkins, "Then the Russians came in a solid little group, which would never separate and

---

[8]Bohm (1959), Bohm and Carmi (1964) and Carmi and Bohm (1964). The papers were submitted on 13 Aug 1962 when Bohm was already at the Birkbeck College and Carmi was at the Yeshiva University in the US. Interview of David Bohm by Lillian Hoddeson on 1981 May 8, Niels Bohr Library & Archives, American Institute of Physics, College Park, MD USA, www.aip.org/history-programs/niels-bohr-library/oral-histories/4513; and Interview of David Bohm by Maurice Wilkins on 1987 January 30, Niels Bohr Library & Archives, American Institute of Physics, College Park, MD USA, www.aip.org/history-programs/niels-bohr-library/oral-histories/32977-7.

[9]Interview of David Bohm by Lillian Hoddeson on 1981 May 8, AIP. Interview of David Bohm by Maurice Wilkins on 1987 January 30, AIP.

never say anything to anybody else. All suspicion and fear. There was [Bogolyubov] and everybody was orbiting around him, all the Russians and satellites, all looking very grim."[10]

Furthermore, these late recollections reveal a feeling of resentment towards David Pines, his former student, regardless of whether it was justified or not. Pines was playing a great role at the conference and looked for talking about the role Bohm had played in the development of the subject of collective coordinates. This made Bohm uncomfortable. "At this Utrecht conference, Pines got up and made a little talk about trying to tell people about what I had done. I felt uneasy about this whole thing, why he should do that or why?" recalled Bohm. To make things worse, Bohm called into question the order of the authorship in the four papers they had worked on together at Princeton: "When I was in Brazil, we had done two papers already on the plasma and the last two were to be done. Pines wrote up the last two, and he said that he would put the last one in his name first, because he had to get a job. I was too busy in Brazil with all sorts of worries to answer him back." As a matter of fact, the fourth paper was signed only by Pines himself. However, this resentment apparently was not so reasonable because Pines' work with Bardeen, Mottelson, and Aage Bohr had highlighted Pines' contribution in this approach and in Utrecht he was possibly trying to balance the credit reminding the audience of Bohm's role in the initial approach. Anyway, regardless of the reasonability of Bohm's feelings, Pines, in the same year as Bohm recorded these recollections in his 1987 interview to Maurice Wilkins, offered a paper—where he recalled their collaboration—to Bohm's 70th anniversary Festschrift. For the historical records their collaboration was never dissociated, as registered in Pines' obituary in the journal *Nature*[11]:

David Pines is best known for his path-forging theory of the collective motion of electrons in metals. He also applied the theory to superconductors, atomic nuclei and neutron stars. Pines developed this theory with David Bohm at Princeton University in New Jersey in the late 1940s. Their work solved a major puzzle in quantum mechanics, in which electrons are described as waves. Early quantum theory predicted the behaviour of electrons in metals, despite not taking into account the repulsion between electrons. Pines and Bohm looked at electrons differently — as quantum plasma, analogous to a gas consisting of charged particles. They realized that electrons move as a group within a metal. Together, the electrons flow back and forth, alternately attracted to the positive ions that make up the crystal lattice and repelled by each other when they get too close to each other. Pines and Bohm showed that these oscillations in the density of electrons are quantized and exist only with a set of allowable energies; they called the oscillations plasmons. Plasmons soften the repulsive forces between individual electrons, thus explaining why the quantum theory worked. Pines and Bohm's seminal papers launched the field of quantum materials. Their collective approach is still used to understand all forms of quantum matter.

[10]Interview of David Bohm by Maurice Wilkins on 1987 January 30, AIP.

[11]Interview of David Bohm by Maurice Wilkins on 1987 January 30, AIP. Obituary: David Pines (1924–2018), Nature, 560, 432, 20 Aug 2018.

5.2 The Quest for the Spiritual—Krishnamurti and the Value of His Thoughts for Society, Science and Individual Beings

Bohm changed his mind not only regarding his view of physics and philosophy of physics. Important changes also happened in his general philosophical beliefs. While these changes would only appear in print years later, we may note a shift between 1957 and 1961. We have already seen the distance he adopted from Communism took place around 1957. We need to add that this frustration with the Soviet Marxism was followed, as we may identify from his letters at the time, with attempts to think otherwise in terms not only of social movements but also concerning the role of ideas and individuals in society compared to the role of economic and social factors. In fact, if we analyze, for instance, one of his longest letters to Melba Philipps explaining new criticisms towards Communism, it is full of references to the role of individuals, human behavior, and political parties as social bodies as explanations for the failures of the Soviet regime. Thus, he lamented Lenin's early disappearance, "There is no doubt, however, that with a leader like Lenin, the worst aspects of the tragedy could have been avoided;" and expressed his frustrations with people in the Communist Parties, "if they really believe what they are saying then one would expect to see Communists behave at least as well toward each other and toward other people as most non-Communists do. Indeed, one would expect more, for their belief in a better society and better humanity should lead them to a higher standard of personal conduct;" and went on to blame the Communist party "for it is essential to remember here that the party systematically encouraged and organized such worship [of party leaders], not only by the character of the propaganda that it put out, but also by its very undemocratic rules, which made criticism of the leaders almost impossible."[12] In hindsight, in the 1987 interview with Maurice Wilkins, Bohm synthesized his intellectual moves. He connected his new views, valuing the world of ideas, with his frustration with the Soviet experience: "They [businessmen, German scientists, Marxists] began to minimize the importance of spirit. When Russia said we've got to establish communism, Lenin's slogan was 'Communism is Soviet power plus electrification', where was the spirit [?]" Still, according to Bohm, "constant stress on the material side would gradually erode the spiritual," and "I was sort of moving toward a broad cosmology and world view which would leave room for something more like spirit."[13]

Still in Bristol, according to Bohm's recollections, his cultural inquietudes had led him to exploit a vast domain of subjects but mainly related to ideas, particularly philosophy and religion. He and Sarah went to the public library to look for books on philosophy, religion, and mysticism. To Maurice Wilkins, in 1987, he recollected "I think I was getting a little depressed there in Bristol for all these reasons." At this

---

[12]Bohm to Melba Philipps, [mid 1956], in Talbot (2017, 182–193).

[13]Interview of David Bohm by Maurice Wilkins on 1987 January 30, Sessions V and VII, Niels Bohr Library & Archives, AIP, College Park, MD USA.

point in the interview the referred reasons concerned the abatement of his interest in physics as it was being practiced, both in Bristol as elsewhere from his experience at the conference in Utrecht. Bohm followed connecting his philosophical inquietudes, quest for new horizons of ideas, and his waning interest in physics[14]:

> I was beginning to feel that physics, you could not base everything on physics, your whole life on physics. I had already been looking into philosophy. I began to look at other things. Sara and I used to go the public library and look at the sections on philosophy, religion and mysticism and so on, looking at some broader issues. The question really was something deeper that would have meaning or value. It seemed that physics did not have as much meaning as I thought it had. It turned out to be not so different from, let us say, business. A businessman does whatever will please his customers and get him money, get him whatever he wants. A lot of physicists seem to be in that boat. They found out what was wanted and did it and hoping thereby to gain various advantages.

Bohm's readings at the time were varied—"I also read Buddhism or oriental philosophy, Indian philosophy, yoga, and probably some of the Christian philosophers"—and these included the mystics George Gurdjieff and Peter Ouspensky. From these early 20th century eastern esotericists, Bohm was particularly impressed by their ideas concerning thoughts, consciousness, and reality: "The idea which they proposed seemed interesting to me which was to become aware of your own thoughts and your reactions and all of life, which are irrational and the source of most of the trouble." Following this thread of interests Bohm came upon the Indian thinker Jiddu Krishnamurti, whose writings and life would touch him deeply. Half chance, half convergence of interests, it was Sarah who recommended this writer at the Bristol library as she thought he was writing things related to Bohm's reflections on quantum mechanics. "Sara picked up this book, 'First and Last [Freedom],' by Krishnamurti. She noticed the phrase, the observer and observed, and thought I talked about it all the time in quantum mechanics, so she thought maybe it had something to do with that and she gave it to me." Bohm began to read the book and his interest was immediately aroused, not only did he read quickly but he ordered more books by the same author. Eventually, Bohm met him personally in June 1961 in London. Over the next 25 years Bohm met Krishnamurti on a regular basis, attended his lectures, participated in his educational initiatives, and together they published a number of books with their dialogues. Bohm's later recollections of their early interaction highlights the subject he and Krishnamurti were touching upon but on different perspectives, the subject which had caught Sarah's attention: "The question of the observer and observed was raised, for example, to say that they were not really separate. I felt from quantum mechanics this must be very significant. He was applying it to the human being himself, saying that the human being as observer was not different from human being as observed." Such a huge interest Bohm developed for

---

[14]Quotations are from Interview of David Bohm by Maurice Wilkins on 1987 January 30, Session VIII, Niels Bohr Library & Archives, AIP, College Park, MD USA, www.aip.org/history-programs/niels-bohr-library/oral-histories/32977-7.

Krishnamurti's thoughts was not limited to finding somebody whose thoughts could lead to analogies with quantum mechanics. Bohm was interested in a further sense[15]:

> Now, this is a very deep point because usually, a human being regards himself as an observer as separate from the observed, even when he is looking at himself. He thinks that he is standing back looking at something inside of himself. But these two are actually one. The confusion that they are separate is the cause of tremendous misery, at least that was saying. I had sort of an intuitive feeling this was right. He was also hinting at something much deeper, some ground, some emptiness in a wholeness ground which everything came, which if we could contact that, then we would sort of rise beyond all these daily problems into a totally different area, where, therefore, we would not be caught in them.

The Indian philosopher Bohm met in 1961 was by then a 65 years old man who had grown up in the care of the leaders of the Theosophy, an esoteric religious movement set in the US in the 19th century, to be the new world religious leader. As a mature man, however, Krishnamurti rejected this role and disbanded the organization which had been set to support the new leader. He did this neither to found a new organization nor to create a new religion. He thought truth could not be taught by any religion or organization. Instead, it was up to each one to improve themselves to find their own interior truth. Krishnamurti's thoughts and life style had attracted the attention of the English writer Aldous Huxley, among many other people in the US and the UK. Apart from his literary talents, Huxley was a humanist, pacifist, interested in parapsychology, mysticism, the theological concept of universalism, and committed to Vedanta, one of the variants of Hindu philosophy It was Huxley who helped to introduce Krishnamurti to a wider audience through the care of a good publisher and the writing of a foreword to Krishnamurti's book which Bohm's had seen. Huxley contrasted the Indian thinker's teachings with all organized religions and beliefs, Communism included, which resonated with Bohm's concerns after his break with Marxism: "We are brought up as believing and practising members of some organization—the Communist or the Christian, the Moslem, the Hindu, the Buddhist, the Freudian." Huxley encapsulated the reach of Krishnamurti's thoughts in a few sentences, which possibly spurred Bohm's current interests[16]:

> In this volume of [...] Krishnamurti, the reader will find a clear contemporary statement of the fundamental human problem, together with an invitation to solve it in the only way in which it can be solved—for and by himself. The collective solutions, to which so many so desperately pin their faith, are never adequate. "To understand the misery and confusion that exist within ourselves, and so in the world, we must first find clarity within ourselves, and that clarity comes about through right thinking. This clarity is not to be organized, for it cannot be exchanged with another. Organized group thought is merely repetitive. Clarity is not the result of verbal assertion, but of intense self-awareness and right thinking. Right thinking is not the outcome of or mere cultivation of the intellect, nor is it conformity to pattern, however worthy and noble. Right thinking comes with self-knowledge. Without understanding yourself you have no basis for thought; without self-knowledge, what you think is not true." [...] "There is hope in men, not in society, not in systems, organized religious systems, but in you and in me."

---

[15] Interview of David Bohm by Maurice Wilkins on 1987 January 30, Sessions VII and VIII, AIP. The full interview is available at https://www.aip.org/history-programs/niels-bohr-library/oral-histories.

[16] All quotations came from Aldous Huxley's foreword to Krishnamurti (1954).

Huxley went on to conclude, in his own words, that "it is through self-knowledge, not through belief in somebody else's symbols, that a man comes to the eternal reality, in which his being is grounded. Belief in the complete adequacy and superlative value of any given symbol system leads not to liberation, but to history, to more of the same old disasters." Neither for Huxley nor for Bohm was there any doubt that the disasters here were references to 20th century tragedies, which included the Soviet experience.

Bohm's rapprochement with Jiddu Krishnamurti requires two clarifications. First, while Bohm came to Krishnamurti's writings after a quest which included mystics such as Gurdjieff and Ouspensky, the characterization of Krishnamurti's teachings is more problematic. Indeed, according to David Moody, who was director of one of the schools run by Krishnamurti, the Oak Grove School, during the process in which Krishamurti distanced himself from the theosophical society which had educated him, he "started to articulate his own, independent message, one steeped in human psychology and not shaped by any esoteric doctrine." However, Krishnamurti's teachings implied the quest for an "ultimate reality" which is "is attainable through immediate intuition, insight, or illumination and in a way differing from ordinary sense perception or ratiocination," to keep with the vernacular meaning of mysticism. In Krishnamurti's own words, "in the religious mind there is that state of silence we have already examined which is not produced by thought but is the outcome of awareness, which is meditation when the meditator is entirely absent."[17]

Second, the relationship observer—observed, which had caught Sarah's eye, was not for Krishnamurti related to quantum mechanics. Bohm was interested in quantum mechanics, not Krishnamurti. For Moody, Krishnamurti's philosophy considered human thoughts as the own source of mankind's infelicities and tragedies. Thus, the main obstacle to free consciousness from biases was "the observer, the thinker, who stands apart from the data of consciousness and feels that he or she is an independent entity who monitors, controls, and decides what course of action to take." He considered this sense of identity an "illusion created by thought, not an independent or objective reality." Thus, "the thinker, the observer, by its very nature, introduces an artificial division in consciousness, when in fact he or she is inseparable from the events under observation." In a nutshell, Krishnamurti often stated that "the thinker is the thought, the observer is the observed." We should thus conclude that if the initial content of the rapprochement between Bohm and Krishnamurti was the relationship between the observer and the observed, the former was thinking of physics, quantum mechanics more precisely, while the latter was concerned with psychology, his specific view about human psychology to be equally precise (Fig. 5.1).[18]

For more than a quarter of century Bohm would maintain an ongoing discussion with Jiddu Krishnamurti on subjects of common interest in addition to regularly lecturing at the Brockwood Park School, in the UK, run by Krishnamurti, as well as regular trips to the Oak Grove School, in Ojai, California, also under the administration of Krishnamurti. Over these years a true literary industry came out of their

---

[17]Moody (2017, 32–35). Mysticism's definition is from Webster's Third New International Dictionary, Merriam-Webster, Springfield, MA (1993). Krishnamurti (1969, 19–20).

[18]Moody (2017, 32–35).

Fig. 5.1 David Bohm and Jiddu Krishnamurti kept lively dialogues for twenty-five years. Krishnamurti and Bohm at Brockwood Park, 1983. *Credits* Photo by Mark Edwards. © Krishnamurti Foundation Trust, Ltd.

interaction. According to David Moody, who was director of the Oak Grove School and who worked closely to Bohm and Krishnamurti for a decade, attending their dialogues, "Over a period of two decades, 144 conversations were recorded between Bohm and Krishnamurti, and many of these dialogues were videotaped as well. Thirty-four were transcribed and edited for publication and appeared in a series of books including the following titles: *The Limits of Thought*; *Truth and Actuality*; *The Wholeness of Life*; *The Ending of Time*; and *The Future of Humanity*." Bohm used to recommend readings and discussions with Krishnamurti to his colleagues and students, for instance, Jeffrey Bub, Donald Schumacher, and Charles Biederman. In some cases, their interactions would not bear fruit and in other cases, such as with Schumacher and Biederman, they would eventually end in clashes.[19]

Bohm's discovery of Krishnamurti's writings, motivated as it was by Sarah's remarks on the observer and observed and its possible relationship to quantum mechanics issues, was in a certain sense an anticipation of cultural trends which would fully flourish ten years later. In fact, as shown by historian David Kaiser, in the book *How Hippies Saved Physics—Science, Counterculture, and the Quantum*

---

[19]Moody (2017, 2). The artist considered that Krishnamurti did not value art, a criticism Bohm accepted. Interview of David Bohm by Maurice Wilkins on 1987 January 30, Session VII, AIP. According to Bohm, "it was unfortunate that Schumacher did not manage to interest Krishnamurti in the language for all around. It would have helped Krishnamurti, and it would have helped Schumacher." Interview of David Bohm by Maurice Wilkins on 1987 April 3, Session XI, AIP.

Revival, research on the foundations of quantum mechanics blossomed in the 1970s driven by many factors including the activity of Californian physicists who were adept to the New Age. Their names included John Clauser, Fritjof Capra, and Nick Herbert. They were particularly motivated by the appearance of Bell's theorem, its quantum entanglement, the possibility of using such phenomena for interactions at a distance, the possibility of quantum mechanics adopting interaction between mind and matter, and by the conceptual connections between these puzzling quantum phenomena and the lessons of Buddhism and Hinduism. Bohm's works on quantum physics and particularly the ideas he would present as a new approach to the quantum phenomena, the approach he would call "Wholeness and Implicate Order," as well as his connections with Jiddu Krishnamurti would fit into these trends. Bohm's ideas were prominently featured in the best-selling book *The Tao of Physics*, written by Fritjof Capra.[20]

In later recollections, in the 1987 interview with Wilkins, Bohm connected almost all the threads we have been discussing. He declared, "I began to feel that nothing was going to solve the human problems that I knew about. Science would not solve it. Politics was clearly was not going to solve it. We had seen after the 20th Congress what had happened to the people who wanted to solve it by socialism." Furthermore, he followed, "Hegelian philosophy would not solve it. It was interesting enough, but it would not solve it. It interested me because it went into this process of thought itself, which I felt was where the trouble is."[21] However, this clear understanding of his own inquietudes and worldview changes is presented only in hindsight, in 1987. From the beginning of the 1960s on it would take years for Bohm to mature and present his ideas either in his research program dedicated to wholeness and order in physics or in his philosophical worldview about dialogue and creativity.

5.3   Attempts to Regain the US Citizenship

The loss of his American citizenship would continue to haunt Bohm in the years he spent in the UK. As early as in 1960, he tried to recover it or even to get a visa, so that he could accept the position that Brandeis University had offered him. His attempts were unsuccessful. He tried again in 1965–1967, with the support of Stirling Colgate, the President of the New Mexico Institute of Mining and Technology, in Socorro. It was Bohm's old friend, Ross Lomanitz, from his times at Berkeley, who mediated the contact with the New Mexico Institute. On April 28, 1965, Bohm thanked to him for helping with "the possibility of exploring how I might return to the U.S. and take a position as professor at your university." Bohm wanted to exploit this possibility while he was "frankly happy in England," and did not "feel a strong urge to return

---

[20]Kaiser's book is Kaiser (2012). See also Herbert (1982) and Capra (1975).

[21]Interview of David Bohm by Maurice Wilkins on 1987 January 30, Session VIII, Niels Bohr Library & Archives, AIP.

to the U.S." Colgate became engaged in the fight as a result of the job offer he had made to Bohm. Again Bohm was unsuccessful.

The background to the first attempt is revealing of the enduring constraints of the Cold War era. In 1960 in London, he was asked by the American Consul about his previous relationship with the Communist Party. Thus, Bohm made a notarized statement on 23 March 1960 about his former links with the Communist Party, and about his current distance from Communist views. Although he had made a notarized statement, Bohm did not intend to make it public. However, this was exactly what the American officials expected of him. Indeed, it would be necessary to demonstrate an active attitude against Marxism, i.e. to make public pronouncements against Communism. Adopting such requirements the US government was in tune with McCarthyism, even in 1960. According to historian Ellen Schrecker, "ex-Communists usually had to purge themselves with HUAC and the FBI before they could work again. The better known among them often had to publish articles in a mass-circulation magazine explaining how they had been duped by the party and describing its evils."[22]

At that stage, Bohm faced a dilemma: either keep his dignity and not recover his American citizenship or recover it, even if it meant losing his dignity. Bohm decided not to pay the price required by the American authorities. His decision is well documented in a letter to Aage Bohr, "It seems that while they are satisfied that I am not a Communist, the McCarran act requires that I prove 'active anti-Communism', e.g. by writing political articles; and this I am not prepared to do." Later, in 1966, Bohm stuck to his decision, as one can see from the letter to Ross Lomanitz, who had been instrumental in recommending him to Colgate, "My principal objection to [publishing something of an 'anticommunist nature'] is that it is not really compatible with dignity. [...] I feel it wrong to say it [his criticisms to Communism] in order to regain American citizenship. For then, I am saying something not mainly because I think it is true, but rather, for some ulterior purpose. It's rather like writing a scientific article in order to impress one's superior, so as to get a better job." It is worth noting that Stirling Colgate understood and supported Bohm's attitude, writing to the US State Department, "He could apply for a visa as an immigrant, and I believe this would require a full demonstration of active opposition to communism with a question on his mind, I am sure, of just how active is active. This question relates, of course, to a sense of personal dignity among his friends and peers."[23]

---

[22]Bohm to Ross Lomanitz, 28 April 1965, (C.42), Bohm Papers. For the date of the "Certificate of Loss of Nationality", see Stirling Colgate folder in Bohm Papers (C.8). "I would like very much to get the question of my US citizenship settled again". Bohm to Stirling Colgate, 28 April 1965, Bohm Papers (C.8). Schrecker (2002, 92).

[23]Bohm to Aage Bohr, November 17, 1960, Aage Bohr Papers, Niels Bohr Archive, Copenhagen. The distinction between declaring not to be Communist and expressing active anti-Communism was not understood by Bohm's first biographer David Peat (1997, pp. 254–255). Peat also asked "Why did he place his rejection of Communism at the end of the Second World War when in fact his letters from Brazil are staunchly pro-Communist?" I think Peat was not very sensitive to the carefully diplomatic manner in which Bohm wrote, in the statement previously cited: "Gradually however, and especially after the war was over, I began to see that ..." Bohm to Ross Lomanitz, 21 Nov 1996, BP (C.42), underlined in the original. Stirling Colgate to George Owen (Deputy Director Visa Office—US State Dept), 4 Nov 4, 1966, BP (C.8).

Bohm's attempt to recover his US citizenship was also related, as we have seen, to his attempts to obtain an academic position in the US, which implies that he was not fully satisfied at Birkbeck in London. The failures to get either the citizenship or a work visa meant his exile from the US would continue. Indeed, in some special cases he could get visiting visas, which he did and visited the US. As we will see later, the citizenship/visa issues would only be solved in 1986, in the twilight of the Cold War, at a time Bohm was no longer considering the possibility of returning to the US.

5.4   Probing the Quanta: Physics and Epistemology

Throughout the 1960s and early 1970s Bohm worked on different topics in physics and epistemology, most of them related to the understanding of quantum mechanics, while simmering at the back of his mind the features of his later approach focusing on the ideas of wholeness and order in physics. We have already made reference to his work with Gideon Carmi on the collective coordinates. The scant reception of this work and his disenchantment with the ways this subject was developing meant that his work with Carmi was a kind of Bohm's swansong in the field he had premiered in the late 1940s.[24] With Aharonov, who had moved to the Yeshiva University in the US, Bohm fought to explain the new quantum phenomenon they had discovered, the Aharonov–Bohm effect, but also entered into a debate with the Soviet physicist Vladimir Fock about the proper understanding of the uncertainty relations to time and energy.[25] Furthermore Bohm kept publishing on the approach led by the French team related to model particles as extended bodies.[26] In the mid-1960s Bohm accepted a new graduate student, Jeffrey Bub, to do his Ph.D. on the foundations of quantum mechanics. Their joint work was at the crossroads of the two most pressing issues in the foundations of quantum mechanics: the extension of quantum mechanics to include new variables and the measurement processes. From the perspective of immediate influence among physicists, Bohm and Bub's papers were the most meaningful of Bohm's works in the 1960s and we will comment on them in a more detailed manner now. Still in the mid-1960s, Bohm published a new textbook, now

---

[24]Bohm and Carmi (1964) and Carmi and Bohm (1964). So far these papers received only four citations, two of them were citations by the authors themselves. To be more precise, however, Bohm would still publish a later paper using collective coordinates (Bohm and Salt 1967), which has 32 citations so far. Source: Web of Science, accessed on 10 Oct 2018.

[25]Aharonov and Bohm (1961a, b, 1962, 1963, and 1964). Unfortunately, we do not have archival material to follow the development of Aharonov and Bohm's collaboration. At the Bohm Papers, deposited at Birkbeck College, there is only one letter, from Aharonov to Bohm, at the folder C.1, dated probably 1961 because after their 1961 papers, reporting to Bohm his first teaching duties at the Yeshiva University: "I am now giving a course on Field-Theory and I find that teaching is not too bad after all. Most surprising of all—even my students enjoy it."

[26]A review of this model and one of the last papers signed by Bohm on the subject was (de Broglie et al. 1963). For an analysis of this research program, see Besson (2018).

dedicated to the principle of relativity. The originality of its pedagogical presentation brought praise. Furthermore, it propitiated a new round of exchanges with Léon Rosenfeld both on the book as well as on the epistemology of quantum mechanics. Bohm also engaged in a long exchange with the American physicist Donald Schumacher on the adequate apprehension of Niels Bohr's thoughts on the interpretation of quantum mechanics, which led to a refinement of Bohm's views. In July 1968, Bohm joined Edward W. Bastin to organize a symposium dedicated to an overview of the state of foundational issues in quantum mechanics and possibilities to overcome it. Bastin was a physicist and mathematician interested in discrete space-time in quantum mechanics and relativity. The proceedings from this symposium gives us a precise snapshot of Bohm's thoughts a little before his shift to a new approach to the quantum issues based on the ideas of wholeness and order. This section is thus devoted to Bohm's works with Bub; the interactions between Bohm and Schumacher, on the one hand, and on Bohm and Rosenfeld, on the other; and finally to the 1968 conference.

5.4.1 Jeffrey Bub, the Measurement Problem, and the Hidden Variables Issue

Jeffrey Bub did his undergraduate studies in physics at the University of Cape Town, South Africa, where he became interested in foundational issues in quantum mechanics through the mathematician and mystic Michael Whiteman. He went to London to study under Karl Popper but at the time Popper was in the US. Bub was advised by G. J. Whitrow to work either with Bohm or Rosenfeld if he wanted to work on foundations of quantum mechanics. The two physicists had opposing views on the foundations of quantum mechanics, a point which Whitrow failed to note. Bub's choice for Bohm, however, was conditioned by prosaic circumstances rather than a philosophical choice, his scholarship was too meagre to support a move to Denmark, and he thought the language would present a problem. He began to work under Bohm in early 1963, at a time Bohm was no longer interested in hidden variables. To follow Bohm in his zigzag attempts to build a general framework for physics based on a discrete space-time structure and on algebraic topology was hardly conducive for someone carrying out doctoral degree research. Bub found his own way through the reading of a paper by Henry Margenau on the measurement problem; subsequently Bohm suggested he "read a paper by Wiener and Siegel, 'The differential space theory of quantum systems,' and consider treating the collapse problem in the framework of a hidden variables theory." More particularly, "Bohm's thought was that one should be able to exploit the Wiener-Siegel 'differential space' approach to quantum mechanics to construct an explicit nonlinear dynamical 'collapse' theory for quantum measurement processes." Thus, with Bub as a student, Bohm returned to deal with the hidden variable issue in a different manner from how he had worked in the early 1950s.[27]

Bub's dissertation led him and Bohm to publish a paper titled "A Proposed Solution of the Measurement Problem in Quantum Mechanics by a Hidden Variable Theory." Their paper was thus a two-pronged work: on the quantum measurement problem and the proposal to solve it with the introduction of hidden variables. Let us begin for the former prong. The choice of the quantum measurement problem revealed Bub's interest in emergent problems in the research on quantum foundations. Before 1963, the description of quantum measurement as a "problem" had not even existed. In fact, while the conceptual roots of the problem date back to the inception of quantum mechanics and the works by Niels Bohr and John von Neumann, it was only in the late 1950s and early 1960s that research in quantum measurement processes gained a certain momentum. Bohm and Bub introduced their paper explaining: "the theory of the process of measurement involves a great many unclear features and unresolved problems, arising mainly because the role of the measuring instrument in the phenomenon of the "collapse" of the wave packet in a quantum mechanical measurement process is obscure. This has been referred to as the *measurement problem* in quantum mechanics."[28] A few years later, Bub still felt it necessary to warn the readers that "I call this a 'problem' in the quantum theory, because the process of measurement in the classical theory is subsumed under the general equation of motion of the theory—measurement has no special status in the theory." As the measurement problem was a novel issue in Bohm's works, the subject deserves a short conceptual digression for the sake of introducing the readers to the historical background of the problem.[29] Below we will return to the second aspect of Bub's work with Bohm, the adoption of a hidden variables approach.

What Bohm and Bub were calling the "quantum measurement problem" was present, albeit implicit, from the inception of quantum theory around 1927. The problem may be summarized as follows: The evolution of the state describing a quantum system, let us say, the electron's spin projection in a given direction, is ruled by Schrödinger equation. This means that the two possible states, spin "up" and spin "down," are expressed in the quantum states and are preserved while time evolves. If one measures this electron spin, one will find it to be either up or down. How did it happen that a state which contains a superposition of two possibilities became just one? Physicists first christened this evolution the "reduction of the wave-packet," an expression which is reminiscent of the wave formulation of quantum theory. A more sophisticated analysis of this process was suggested by Niels Bohr, assuming that measurements require macroscopic devices while appealing to the

---

[27]Bub (1997, xi–xii). Margenau's paper was Margenau (1963) and the papers by Norbert Wiener and Armand Siegel were Wiener and Siegel (1953, 1955), Siegel and Wiener (1956), Bohm and Bub (1966a, b).

[28]Bohm and Bub (1966a, 454) and Freire Junior (2015, 152).

[29]Bub (1971, 65). In the 1951 hidden variables interpretation Bohm dealt with the problem introducing the additional variables both in the system and in the measurement device and coupling them. In the two next paragraphs I am following my Quantum Dissidents' book: (Freire Junior 2015, 142–143).

complementarity view suggested by himself. Bohr suggested that such devices had to be treated within the framework established by classical physics, not because one could not treat them from a quantum point of view, but because they had to be treated classically so that measurement results could be compared to those of other researchers. As communication is a requirement for objectivity, and communication requires ordinary language refined by concepts from classical physics (e.g., concepts described in words such as "work," "force," etc.), the classical treatment of measurement devices is a condition for preserving objectivity in scientific research.[30]

The other major solution to the measurement problem was suggested by John von Neumann as part of his work to lay rigorous mathematical foundations for the mathematical formalism of quantum physics. He began to work on this subject just after the elaboration of the mathematical formalism of the quantum theory, around 1925–1927. In 1927 he wrote a trilogy of papers which would be the basis for his book *Mathematische Grundlagen der Quantenmechanik*, published in 1932 in German and only translated into English in 1955. Considering that the "transformation theory," formulated by Dirac and independent contributions by Pascual Jordan and Fritz London, were the "definitive form" of quantum mechanics, von Neumann departed from it due to its lack of mathematical rigor. According to him, in the opening of his book, "it should be emphasized that the correct structure need not consist in a mathematical refinement and explanation of the Dirac method, but rather that it requires a procedure differing from the very beginning, namely, the reliance on the Hilbert theory of operators." Von Neumann not only based his presentation on the mathematical structure of Hilbert vector spaces and Hermitian operators, but also extended it beyond its "classical limits." From then on, matrix mechanics, wave mechanics, and transformation theory should be considered as manifestations of Hilbert space vectors.

Von Neumann's and Jordan's work has been dissected by historians Anthony Duncan and Michael Janssen, who argued, "So, rather than following the Jordan-Dirac approach and looking for ways to mend its mathematical shortcomings, von Neumann, as indicated in the passage from his 1932 book quoted above, adopted an entirely new approach. He generalized Hilbert's spectral theory of operators to provide a formalism for quantum mechanics that is very different from the one proposed by Jordan and Dirac." In the search for the consistency of his mathematical scheme von Neumann used this formalism to deal with measurement in quantum physics and he diverged from Bohr's solution. The milestone in von Neumann's treatment of measurement was the introduction of a distinction between two kinds of time evolution of quantum states. The first one, "discontinuous, non-causal and instantaneously acting experiments or measurements," occurs during the measurement processes and is governed by a special type of operators, the projection operator. The second one, "continuous and causal," is ruled by the Schrödinger equation. In addition, von Neumann treated measuring devices quantum mechanically, instead of treating them classically as suggested by Bohr. This choice leads to the transfer of

---

[30]For a standard, comprehensive description of complementarity, see Bohr's (1949) report of his discussions with Einstein.

the singular superposition of quantum states from the system under scrutiny to the combination of system and measuring apparatus. In mathematical terms, this transfer is represented by the inner product between the two Hilbert vectors, one related to the system and the other related to the measurement device. As no such measurement device described by such a bizarre superposition has been seen, it raises the questions: how, where, and when does this superposition become a vector with just one component, which is an eigenstate of the physical property of interest? After all, what we obtain after measurements is related to magnitudes and their statistical frequencies rather than to superposition of vectors. Von Neumann solved the problem appealing to the distinction between the two kinds of evolution of quantum states and the role of the cognizant subject, that is, the individual observer. He recalled the general epistemological view that in any measurement there ultimately is a moment in which "we must say: and this is perceived by the observer," and framed his answer in the requirement of the psycho-physical parallelism, which means "it must be possible so to describe the extra-physical process of the subjective perception as if it were in reality in the physical world," i.e., "to assign to its parts equivalent physical processes in the objective environment, in ordinary space."[31]

The story of the study of the quantum measurement processes from Bohr's and von Neumann's initial views to the debates in early 1960s would take us beyond the scope of this book. It is enough to note that Bub and Bohm were not satisfied with either of these approaches and looked to change quantum mechanics in order to get an explanation for the "reduction" or "projection" deriving it from the same mathematical formalism ruling the time evolution of the quantum states. In a more precise manner, they went "on to propose a new deterministic equation of motion, describing a kind of coupling of the measuring instrument to the observed system that explains in detail how the wave packet is 'reduced' during a measurement in a continuous and causally determined way." Furthermore, they criticized the proposal formulated by the Italian physicists Daneri, Loinger, and Prosperi, and strongly supported by Rosenfeld to explain the collapse appealing to the notion of an "amplification" occurring during the coupling between the system and measurement apparatus. Bohm and Bub argued that such an explanation could not cope with the situation of collapse happening in the state describing individual systems. According to them: "The fact that interference between different parts of the wave function, corresponding to different values for the observable measured, is effectively destroyed in the amplification process does not explain why the wave function of the individual total system 'condenses' onto one and only one of the component noninterfering wave packets that are produced in the interaction with the measuring apparatus. Thus, the question of the behavior of an *individual total* system is avoided." The proposal made by the Italian physicists was also criticized by a number of different physicists, including Eugene Wigner, and was eventually abandoned. However, the debate over this proposal contributed to shake up the hegemony of the orthodox views, such as those of Bohr and von Neumann, in the field of foundations of quantum mechanics. Still, Bohm and Bub's proposal were among the first attempts to change Schrödinger equation in order to

---

[31] von Neumann (1955, pp. ix, 419), Duncan and Janssen (2013, 194).

solve the measurement problem. Twenty years later these attempts led to the spontaneous collapse approach, and it is still in debate, at least in the circle of physicists working on the foundations of quantum mechanics.[32]

Now let us come back to see when and how hidden variables reappeared in Bohm's works. Bohm had suggested that Bub take a look at the mathematical approach called differential spaces, developed by Norbert Wiener and Armand Siegel, to deal with the measurement problem. There is a hidden history in this suggestion, which certainly was unknown to Bub and possibly only partially known to Bohm. Indeed, Wiener had been motivated by Bohm's early work on hidden variables while he and Siegel developed their work in a different way. According Wiener, "I have been tremendously influenced in my thinking by my conversations and correspondence with Mr. Gabor and Mr. Rothstein, and by reading a sequence of two papers [...] which appeared this January under the authorship of David Bohm."[33]

The work by Bub and Bohm was, however, distinct from Bohm's early work in a number of aspects. First because, as we have seen, it was intended to solve the measurement problem through the introduction of "a deterministic equation of motion [...] which couples the measuring instrument to the observed system and so explicitly incorporates the role of the measuring instrument into the dynamics of the theory." Second, because this has "potential contrast with quantum mechanics." As their theory was a change in the standard quantum mechanics they suggested, after averaging over the hidden parameters, a time to recover the quantum statistical results as a special case. They suggested a threshold, a relaxation time of the order of 10^{-13} s, below which there would appear conflicts with quantum mechanical predictions. As far as I know this was the only time Bohm suggested a figure to contrast hidden variables with quantum mechanics. In the 1950s, he had just begun to speculate that changes in his model could produce different predictions in the domain of the size of an atomic nucleus, but he did not carry out the promised changes. Immediately after Bohm and Bub's proposal, the Harvard experimentalist Costas Papaliolios seized the opportunity to test it. Papaliolios successively measured linear polarization of photons emitted from a tungsten-ribbon filament lamp. The measurements were carried out within time intervals lower than the threshold suggested by Bohm and Bub, and he found their theory untenable.[34]

Bohm was notified of the result before its publication and reduced the reach of this experiment, "I regard our 'theory' largely as something that is useful for refuting von Neumann's proof that there are no hidden variables. I would not regard it as a definitive theory, on which predictions of experimental results could be made." In

---

[32]Bohm and Bub (1966a, 453–459). The debate on the measurement problem and on the Daneri-Loinger-Prosperi work, including the participation of Wigner, Rosenfeld, Bohm, Bub, Daneri, Loinger, Prosperi, and Klaus Tausk, among others, is analyzed in (Freire Junior 2015, 141–195). Bub's follow-up in the debate with the Italian physicists is (Bub 1968a). On the spontaneous collapse approach, see Frigg (2009).

[33]Norbert Wiener, "Paper to be presented on May 3 [1952] before the American Physical Society by Norbert Wiener," [7 pp, unpublished,] Box 29C, folder 678, Norbert Wiener Papers, Institute Archive, MIT, Cambridge, MA.

[34]Bohm and Bub (1966a, 454), Papaliolios (1967).

addition, he admitted that the time suggested "is just a guess," and not a consequence of their theory.[35] Papaliolios conceded that "the primary purpose of [Bohm-Bub's] paper was to demonstrate, by means of an explicit theory, how one can circumvent von Neumann's proof," but emphasized the role that experimental predictions and real experiments, such as the one he had carried out, should play a role in the choice between theories in the foundations of quantum physics. In the same year, another physicist, Jerald Tutsch, from the University of Wisconsin, followed the debate "In the paper by Bohm and Bub an ad hoc randomization time of h/kT for the hidden variables is suggested. An experiment on the polarization of light by Papaliolios showed this estimate to be too large by a factor of at least 75." His conclusions, along Bohm's lines was: "Since the collapse time is connected to the Bohm—Bub theory in a fundamental way and the randomization time is not, the outcome of the experiment is of little significance."[36]

As Bohm had written to Papaliolios, Bohm and Bub's main motivation was to demonstrate flaws in von Neumann's proof against the possibility of introducing hidden variables in quantum mechanics. Bohm and Bub used their model as a counter-example to the proof, a strategy similar to that used by Bohm in his 1952 papers on the causal interpretation. They published their paper at a moment when the debate over von Neumann's proof was heating up. In 1963 Josef-Maria Jauch and Constantin Piron published a paper trying to reinforce von Neumann's proof. In 1966, in the same issue of the *Reviews of Modern Physics* where Bohm and Bub published their paper, Bell also published one of his two seminal papers on hidden variables. He identified an assumption in von Neumann's proof that seemed to him to be unacceptable because while quantum mechanics satisfies this assumption it is not reasonable to require that any alternative theory have the same property. This assumption was that "any real linear combination of any two Hermitian operators represents an observable, and the same linear combination of expectation values is the expectation value of the combination." Bell also addressed the Jauch-Piron work on the same grounds. In a follow-up paper Bohm and Bub made a devastating criticism of Jauch and Piron's paper, stating "while the assumptions of Jauch and Piron are in fact weaker than those of von Neumann, the net result is that they actually prove nothing new at all." In more technical terms, Bohm and Bub's reasons were that "the conclusions of Jauch and Piron concerning the nonexistence of hidden variables are indeed seen to follow from a false assumption; i.e., that the impossibility of propositions that describe simultaneously the results of measurements of two noncommuting observables is an 'empirical fact.'" Still according them, "actually, it is shown that this assumption follows if and only if one first assumes what the authors set out to prove; i.e., that the current linguistic structure of quantum mechanics is the only one that can be used correctly to describe the empirical facts underlying the theory." Bohm and Bub's paper was followed by at least three papers debating the subject. It was a

[35]Costas Papaliolios to David Bohm, 17 February 1967 and 20 March 1967; Bohm to Papaliolios, 1 March 1967, 2 March 1967, 11 May 1967. Papaliolios Papers, Accession 14,811, Boxes 23, folder "Hidden variables," and 10, folder "Bohm letters," respectively; Harvard University Archives.
[36]Tutsch (1968).

coincidence that these papers by Bell and by Bohm and Bub were published in the same issue of the same journal. The connections between Bell's and Bohm's works are strong as Bell had been motivated to work on the hidden variables issue, since the early 1950s, by Bohm's 1952 paper. Bohm and Bub were well aware of Bell's paper and their work was presented dialoguing with Bell's paper. However, while Bohm and Bub dialogued with Bell's 1966 paper, they did not cite the second seminal Bell's paper, where what now we know as Bell's theorem appeared. According to Jeffrey Bub, Bell's papers and his theorem were discussed by Bohm and him, but Bub also recalls that only later, while in Minnesota, did he fully realize the implications of Bell's theorem. At the end of this chapter we will come back to this second Bell's paper to comment on its influence on Bohm's ulterior work.[37]

From mid-1966, after concluding his Ph.D. at the Birkbeck, Jeffrey Bub began to teach at the University of Minnesota. In the following three years he and Bohm maintained a huge and lively correspondence about quantum mechanics—measurement problem and hidden variables—and philosophy. Most of the exchanges concerned the two works Bub was writing as the follow-up of his dissertation and they were eventually published in 1968 and 1969. In these interactions Bub, helped by Donald Schumacher, to whom we will also return, pushed Bohm for a deeper interpretation of Niels Bohr's complementarity view as well as an enhanced explanation of Bohm's own earlier work on the causal interpretation. On Bohr's views, Bub emphasized the connections between the use of classical concepts and the requirement of communication of experimental results: "the 'feature of wholeness' is introduced via an argument which involves the assumption that unambiguously communicable information about what is observed is necessarily described in terms of the concepts of classical physics, or refinements of classical concepts." In addition, Bub thought the feature of wholeness could be preserved while dropping the dependence on the classical concepts.

On Bohm's early work on hidden variables, Bub did not see its features as leading to the restoration of classical determinism and criticized the term itself as the term was associated to von Neumann's proof against hidden variables, which Bub thought had no relevance for Bohm's 1952 proposal. Thus, in a surprising conclusion, at least considering the harsh debates of the previous years, Bub suggested a change in terminology in order to reconcile Bohm's and Bohr's works: "the term [hidden variables] should not be used as a label for the theories considered by Bohm and other workers in this field. Such theories could be regarded as fundamentally compatible with the original Copenhagen interpretation of the quantum theory, as expressed by Bohr." Reflecting the times of political and cultural upheavals, the late 1960s, Bub strived to distinguish a conservative from a revolutionary Niels Bohr[38]:

> From this point of view, Bohr's assumption about the significance of classical concepts may be regarded as the conservative aspect of his interpretation of the quantum theory, and his idea of 'wholeness' as the essence of this interpretation. The hidden variable approach is

---

[37]Bell (1966), Jauch and Piron (1963), Bohm and Bub (1966b, 470). Talk with Jeffrey Bub, 22 May 2002, American Institute of Physics, College Park, MD.

[38]The two papers were Bub (1968b, 1969). Bub (1968a: 185, 1968b: 186, 1969: 101).

an attempt to extend Bohr's basic insight into the revolutionary significance of the quantum theory, and not a conflicting proposal to restore classical determinism. What is rejected is Bohr's 'conservative thesis'; what is retained, in a new sense, is Bohr's 'revolutionary thesis.'

Bohm reacted favorably to this evolution of Bub's thoughts, dedicating more than 500 pages of letters to Bub. "What you say about Bohr […] is very pertinent. It is true that he saw that the quantum implies the wholeness of physical phenomena. […] The key to really seeing wholeness in nature is both to enrich the common language and to develop new forms of relevant mathematics," Bohm wrote to Bub on February 22, 1967. Later, on July 20, 1967, he followed "what is Bohr's point of view? It is that the question is basically epistemological." On October 16, 1967, Bohm dedicated the full 11-page letter to distinguish Bohr's and von Neumann's views on quantum mechanics. And still, on November 5, 1968, "perhaps the words 'hidden parameters' or 'hidden variables' should now be dropped altogether. Instead we could talk of 'an extension of a given description, to include further parameters that are contingent in the context of the original description.'"[39]

Bohm understood Bohr's argument on the role played by communication considerations but could not agree on this. Thus in his later writings we will see a more faithful presentation of Bohr's complementarity but also an emphasis on the distinction between epistemology and ontology when the interpretation of quantum mechanics is concerned. At a certain point however, Bohm's and Bub's views began to differ. Bub was considering von Neumann's axiomatic treatment, and his later work on quantum logic, in a more favorable light, while Bohm could not agree with this, as he had always taken distance from more abstract and formal approaches. Thus, on December 5, 1968, he wrote to Bub, "I am now very much struck by your statement that you are in disharmony of opposing judgments of relevance (i.e. those of the group following von Neumann and that you find in my work) […] in some sense, you are trying to 'reconcile' my relevance judgments with those of V. Neumann, Jauch and Piron, Gudder, et al. Since this cannot be done, you are the 'problem' of trying to achieve the impossible. […] It is clear that v. Neumann and his followers rule out Bohr as irrelevant, not by saying so explicitly, but rather, by simply not mentioning him at all."[40] These differences were in anticipation of Bub's later intellectual interests which shifted to quantum logic. Bub's views ultimately evolved for a kind of reconciliation between the two themes he had worked through his life: hidden variables and quantum logic. In 1998 he won the prestigious Lakatos Award with the book *Interpreting the Quantum World* where this reconciliation is presented (Figs. 5.2 and 5.3).[41]

---

[39]These Bohm's letters are at Bohm Papers, Folders C.131, C.132, C.133, and C.135, respectively, Birkbeck College, London.

[40]Bohm to Jeffrey Bub, 5 Dec 1968. C.135, Bohm Papers. Bub's interest was mainly focused on von Neumann and George Birkhoff's study of the logic intrinsic to quantum mechanics. An early exposition of Bub's view is in Suppe (1974, 406–408).

[41]Bub (1997, p. xiii).

Fig. 5.2 Basil Hiley, who
became Bohm's assistant for
decades, and David Bohm in
the 1970s. *Credits* Courtesy
of Basil Hiley

Fig. 5.3 Jeffrey Bub, in the
mid-1960s, first student of
Bohm's whose doctoral
subject was entirely
dedicated to the foundations
of quantum mechanics.
Credits Courtesy of J. Bub

5.4.2 *Relativity Textbook, Rosenfeld, Schumacher, and Philosophical Debates*

After his first years at Birkbeck, Bohm wrote a new physics textbook, which was
titled "The special theory of relativity," inspired by his teaching experience. In addition to being written with notable clarity, the book presented at least two novelties
compared to other relativity textbooks available at the time. Bohm spends almost the
first third of the book exploiting pre-Einstenian relativity, including Lorentz's theory
of electron. Thus the reader is introduced to all the conceptual issues which paved
the way to the creation of the special principle of relativity by Einstein. Then the
reader is shown how departing from this principle, that is, that all physical laws must
have the same form in any systems of coordinates moving in uniform translation in

relation to other system, one may derive all Einstein's relativistic formulas, including his famous $E = mc^2$. In addition, the book has an appendix entitled "Physics and Perception" which shows it is illusory to think that Newtonian concepts agree with everyday perceptual experience. Following the psychologist Jean Piaget's investigations, Bohm argues "that many of our 'common sense' ideas are as inadequate and confused when applied to the field of our perceptions as they are in that of relativistic physics." Furthermore, "there seems to be a remarkable analogy between the relativistic notion of the universe as a structure of events and processes [...] and the way in which we actually perceive the world through the abstraction of invariant relationships in the events and processes involved in our immediate contacts with this world." While dialoguing with psychology Bohm was led to conclude that "science is *mainly* a way of extending our perceptual contact with the world, rather than of accumulating knowledge about it."[42] Finally, Bohm only used a mathematical treatment to convey the finer understanding of relativity exploiting as much as he could a conceptual treatment. The mathematical treatment led to the introduction of Minkowski space but it did not use tensorial calculus.

The book was well received even before its publication. According the publisher, he could not "recall ever reading a more ecstatic review of a manuscript," and one of the referees opened his report exclaiming: "how wonderful it is, in a life of many manuscripts, to receive a book from a fluent scholar! This is a most competent, provocative, and engaging effort, and I hope it will soon appear for us to use." Reviews were along the same lines, while not unanimous, but the most surprising, for Bohm, came from Rosenfeld, who had been one of Bohm's harshest critics. Bohm wrote to him, after receiving a typescript of his review, "it was indeed far more laudatory than I would have expected from anyone at all, and especially from you." Indeed, Rosenfeld opened his review writing: "I regard this book (which give us much more than its title suggests) as one of the most remarkable productions in many years in the domain of philosophy of science on the one hand, and of higher education on the other."[43] Rosenfeld's reaction encouraged Bohm to resume with him discussions about the interpretation of quantum mechanics. These exchanges lasted till the end of 1968 but through this correspondence the old differences reemerged albeit with different nuances.

Bohm used a subterfuge to engage Rosenfeld in a continuity of correspondence now on the foundations of quantum mechanics. Bohm's strategy was to use a letter to Loinger, which dealt with the measurement problem, to attract Loinger and Rosenfeld

---

[42]Bohm (1965a, xi).

[43]Robert Worth to David Bohm, 26 Nov 1963; Anonymous report to R. Worth, 27 Nov 1963, Bohm Papers, Folder C.64. Bohm to Rosenfeld, 8 Sep 1965, with an enclosed review, Rosenfeld Papers. For other reviews, see C. W. Kilmister [Supplement to Nature, Dec 4 1965, pp. 986–987], who considered it a "most satisfactory book for background reading for anyone learning special relativity," and G. C. Vittie [Science, 149(3682), 415–416, 1965], who criticized the misleading ideas conveyed by the lack of the tensor calculus. Vittie prodded Bohm titling his review "A 'Three Plus One' Dimensional Treatment." However, Vittie misinterpreted Bohm when he wrote Bohm's book implied to consider the "world of common-sense" as a "guide to the basic ideas of physics;" indeed our analysis suggests quite the opposite.

to a general philosophical discussion about Niels Bohr's views. Indeed, he moved the subject from the criticisms on the insufficiencies of the DLP model to a philosophical discussion about complementarity. In addition, in the same letter, he suggested to Loinger forward his letter to Rosenfeld once Rosenfeld had strongly supported the DLP approach as a completion of complementarity. Thus Bohm began by saying that "Bohr has been giving top priority to the role of language," then called Aage Petersen as testimony to state that, "as Petersen says, Bohr was fond of emphasizing that 'man lives suspended in language.' Therefore, to Bohr, the principle of complementarity is primarily a linguistic question." Bohm thus concluded Bohr's stance was not "primarily physical, but rather epistemological." So far so good, or at least one would think Rosenfeld would not be concerned had Bohm stopped here. However, Bohm went on to say "I think that Rosenfeld does a disservice both to Bohr and to his own dialectical materialist views, by failing to recognize that Bohr is not a materialist (though he may perhaps be, in certain ways, a dialectician)." Furthermore, Bohm revealed "indeed, I learned in conversations with Petersen that in his early life, Bohr was strongly influenced by Kierkegaard." And concluded, "One can in fact see the influence of existentialism in the principle of complementarity. Man is somehow an individual and yet he must be in indivisible union with God. [...] Similarly, in physics, Bohr takes subject and object as being indissolubly united by the quantum that 'connects' them."[44]

Bohm was entering muddy waters with this letter. He was not only challenging the coherence of Rosenfeld's philosophy. In fact, in the mid-1960s, after Bohr's death on November 18, 1962, the battle for the intellectual legacy of the Danish physicist had begun. Part of this battle was being fought in Copenhagen itself as Aage Petersen, once Bohr's assistant, was advancing, as part of his doctoral research in philosophy, his interpretation of the relationship between Bohr's views and the Western philosophical tradition. Rosenfeld could not disagree more with Bohm's stances. "I have shown the letter to Aage Bohr, who was as startled as myself. We do not, of course, take very seriously this game of putting labels with various 'isms' upon Bohr," began Rosenfeld. "I feel I ought to write to you about the facts of the case, because I suspect that you have been badly misinformed by Petersen,". Rosenfeld went on to argue that while Bohr had read the Danish theologian and existentialist philosopher, admired his literary style, he did not follow his thoughts. Rosenfeld maintained that there was no compatibility between the "extreme irrational subjectivism" and Bohr's objective rationalism. "In particular, the idea of mixing up God with Bohr's thinking appears most incongruous to those who have known Bohr intimately and who remember his familiar saying: 'God is just a word of three letters',," wrote Rosenfeld. Finally, he returned Bohm's prodding in a paragraph whose overtones recall Bohm's earlier commitment to Soviet Marxism: "Incidentally I am afraid that you have also misunderstood what I have said about the relation of Bohr to materialism. I certainly have never played the game of putting a materialist label on him. I leave that to the high-priests of Marxist-Leninist theology." Rosenfeld thus concluded stating, "all I wanted to suggest is that Bohr's thinking, as well as

---

[44]Bohm to A. Loinger, 10 Nov 1966, C.130, Bohm Papers.

Einstein's thinking in his creative period, occupies a natural place in a dialectic development initiated by the materialistic philosophy of the French school of physics and chemistry [...] at the end of the 18th century."[45]

After these initial shots Bohm and Rosenfeld, over the next two years, exchanged their correspondence in a softer manner while maintaining their different point of views. Bohm conceded on a few points, "I must say that one hears different accounts of Niels Bohr's philosophy. However, I did not wish to suggest that Bohr took religion seriously." Following this letter Bohm sent a second one trying to appease relationships not only with Rosenfeld but also with Aage Bohr, the son of Niels Bohr with whom Bohm had developed a friendly relationship: "assure him that I had the very highest esteem for his father. If I am deeply moved by the questions that moved Niels Bohr, this may cause me to be a bit sharp with his views at times." Then Bohm added, "but I am sure he would have preferred such a response to the indifference with which physicists now so generally respond to these questions." Rosenfeld praised Bohm's manuscript entitled "Creativity" and advised Bohm to publish it in *Nature* through the good will of John Maddox, editor of the journal, who had served on Rosenfeld's staff while he had worked in Manchester. Furthermore, Bohm engaged Rosenfeld in the organization of a symposium entitled "Quantum Theory and Beyond" he was organizing at Cambridge with Ted Bastin. While Bohm and Bastin appeared as its "proposers," Rosenfeld figured as its "sponsor," but he could not attend it.[46]

Bohm's correspondence with Rosenfeld also dealt with Donald Schumacher, an American physicist who was interested in working on Niels Bohr's ideas on complementarity. Bohm asked Rosenfeld's support to obtain a fellowship to bring Schumacher for a stay at Birkbeck. On December 8, 1966, Bohm asked for Rosenfeld's views on Schumacher's stances: "I have been in correspondence with a man named Schumacher in Cornell University. He seems a very intelligent fellow, though because his style of writing is modelled on that of Niels Bohr, he is often almost incomprehensible to me." Then Bohm presented the content of Schumacher's ideas, "I have gotten from him the impression that Bohr regards the algorithms of quantum theory as a sort of 'metalanguage,' in terms of which one makes statements about the 'ordinary' language of physics, i.e., classical physics. In other words, according to Bohr, all the 'objective' features of the world are those that can be precisely communicated, because they are in the language of classical physics." Rosenfeld was skeptical about Schumacher's prospects: "I am afraid that he has been infected at some stage by the jargon of modern logicians which is ideally suited to make the simplest things sound

---

[45]Rosenfeld to Bohm, 6 Dec 1966, Rosenfeld Papers. Petersen's dissertation was published as (Petersen 1968). For a presentation of Rosenfeld's views as well as on the case with Petersen's dissertation, see (Jacobsen 2012); see in particular p. 270, footnote 54, on the relationship between Rosenfeld and Petersen.

[46]Bohm to Rosenfeld, 8 Dec 1966, 13 Dec 1966, 15 May 1967; Rosenfeld to Bohm, 18 May 1967, 17 Jan 1968, 23 July 1968. Rosenfeld Papers. "Creativity" was not published in *Nature* in the end, it was too long, instead it appeared in the new journal *Leonardo* (Bohm 1968); it was the departure for a long research by Bohm which appeared as a book as (Bohm and Peat 1987) and, posthumously, in 1996, as (Bohm 1996). The symposium "Quantum Theory and Beyond" was the basis for the volume (Bastin 1971).

terribly complicated." The initial interaction between Schumacher and Bohm was marked by sharp differences as the former considered randomness the central lesson from quantum physics and considered Bohr's thoughts "unclear but consistent," which Bohm disagreed with. However, their interaction evolved positively due to Schumacher's emphasis on the role of language in Bohr's reasoning. Bohm did not agree with this, considering it epistemology rather than physics, but was very interested in a deeper and diversified analysis of Bohr's thoughts. Ultimately, we can say that Bohm benefitted from discussions with Schumacher, Petersen, Bub, Hiley, and even Rosenfeld, to form his own view. As he reported to Rosenfeld, on May 26, 1968, "I have some new ideas that bring me closer to Bohr's position (though there are still serious divergences)." Bohm and Schumacher began to collaborate on a joint paper presenting Bohr's and Einstein's divergences as a failure of communication.[47]

The joint paper, Schumacher and Bohm, while written as a first draft, was never published because rifts began to appear. Bohm reported that Schumacher began to disagree with the paper content and then had a psychological breakdown and ended up in hospital with a diagnosis of schizophrenia. According to Bohm's recollections, "the trouble was that we wrote something on Einstein and Bohr together, Schumacher and I, and Schumacher then broke down. He was rather unstable and he broke down into paranoid schizophrenia and he was hospitalized, and before that, he was getting all sorts of disturbed reactions to this. He went against it and he began to change it into an incomprehensible thing, and it became impossible to do anything with it." Bohm also reported the high esteem he had for Schumacher and his work, recalling: "During this period which I found somewhat depressing as far as physics was concerned, we got a student, Donald Schumacher, was quite brilliant. He came from America, but he was not very stable mentally. He took a great interest in Niels Bohr. He really studied Niels Bohr and he had some insights into Niels Bohr." About the influence of Schumacher's work on Bohm's own views about Bohr's thoughts, Bohm was rather emphatic: "The point is, I had not really understood Bohr, but I had sort of seen him the way I wanted to see him. Because he was so hard to understand, I sort of began to read my own view in there. This fellow Schumacher had some insight which made it much more clear what Bohr was about. What he said could be summed up by saying that the form of the experimental conditions and the meaning or content of the results are a whole, not further analyzable. This question of the observer and the observed, which is fascinated by quantum mechanics, is very hard to put consistently without confusion."[48]

---

[47]Bohm to Rosenfeld, 08 Dec 1966, 4 Oct 1967, and 26 May 1968; Rosenfeld to Bohm, 19 Jan 1967, Rosenfeld Papers. Bohm to Donald Schumacher, 24 Oct 1966, 23 Nov 1966, and 20 Dec 1966, Bohm Papers, Folder C.130.

[48]The draft is D. Bohm and D. Schumacher, "On the failure of communication between Bohr and Einstein," typescript, 9 pp, Folder B.44, Bohm Papers. Interview of David Bohm by Maurice Wilkins on 1987 February 27, Session IX, Niels Bohr Library & Archives, AIP, College Park, MD USA, www.aip.org/history-programs/niels-bohr-library/oral-histories/32977-9; and Interview of David Bohm by Maurice Wilkins on 1987 April 3, Session XI, Niels Bohr Library & Archives, AIP, College Park, MD USA, www.aip.org/history-programs/niels-bohr-library/oral-histories/32977-11. Accessed on 18 Oct 2018.

Schumacher however had a stronger view about the context surrounding his divergence with Bohm and his breakdown. He reported to Henry Horak, who had been his masters advisor at the University of Kansas, on May 9, 1971: "Thanks for your letter of some time ago. It cheered me up quite a bit. I was in the hospital at the time. I'd had a nervous breakdown after working a little hard on a short book and they kept me in that zoo for about two months. I've just about given up science. Too many disappointments... I've packed in working for Bohm about a year and a half ago." Horak added that on October 4, 1972, he received another letter from Schumacher, with a collection of 6 papers enclosed, stating "the last two papers are of special importance both from the standpoint of fundamental physics and from that of logic and philosophy ... (and) it will make the 'failure of communication' with Bohm more clear and will make arguments which I was not able to make at the time–but which I found were made previously by Wittgenstein, and not understood." Later Schumacher published a paper which focused on the analysis of Bohr's thoughts, with a Wittgenstein's epigraph, which is reminiscent of this project. No reference to Bohm nor their discussion appeared in this paper.[49]

Two of the conferences Bohm participated in the late 1960s shed lights on the philosophical and scientific issues he was dealing with. The first one was a symposium in Illinois organized by the philosopher Frederick Suppe, in 1969, while the second happened in Cambridge organized by Bohm and Ted Bastin, in 1968. Let us recall Bohm had written in his Relativity textbook that "science is mainly a way of extending our perceptual contact with the world, rather than of accumulating knowledge about it." He took Suppe's invitation as an opportunity to clarify and counter balance this statement. In fact, his talk was titled "Science as Perception-Communication," no doubt bringing a social dimension to the production of scientific knowledge and preventing an idealist reading of his earlier statement. Bohm focused his talk in criticizing the ideas of "normal science" and "incommensurability" among theories, which had been introduced by Thomas Kuhn in his book *The Structure of Scientific Revolutions*. Bohm's point was that even in periods considered by Kuhn as being normal science there were important conceptual changes. "According to him, "the fact that progress in what is called a quiet period of normal science often involves such fundamental change, the relevance of which is overlooked, can be illustrate by considering certain features of the development of the quantum theory." He illustrated his argument with the differences between Bohr's approach to quantum mechanics and von Neumann's presentation of the same theory. He also called an example from the 19th century physics: "it was widely believed in the nineteenth century that Newtonian dynamics and Hamilton-Jacobi wave theory of dynamics were 'essentially the same.' Nevertheless, we can now see that the difference between

---

[49]Schumacher's letters are cited in Henry Horak, "Philosophy, astrophysics, and time's arrow. Donald Schumacher," available at https://physics.ku.edu/astronomy/history/horak/17, accessed on 18 Oct 2018. Horak and Schumacher had previously co-authored the paper (Schumacher and Horak 1962). On Horak, see his obituary written by K. Horak, in *Physics Today*, 16 Jan 2013, https://doi.org/10.1063/pt.4.1765. The reported six papers were not published, at least in technical journals as far as I could check. The published paper is Schumacher (1974).

'wave dynamics' and 'particle dynamics' was potentially of very great relevance in the sense that the former can lead in a natural way to quantum theory, while the latter cannot."[50]

Bohm's paper stirred a wide debate, from which we will take Kuhn's, who was at the audience, reaction. Kuhn opened his statement acknowledging difficulties with the term commensurability—"It is a term I have used, although it is not one I have been very fond of"—and went on to consider that Bohm also had not properly dealt with the issue of communication across different theories, an adequate criticism considering the text of Bohm's paper. In a certain moment, however, Kuhn revealed he considered that "it is my impression that one of the greatest difficulties faced by people who are concerned to revise the Bohr interpretation is that none of the problems that emerge for them makes any contact whatsoever with the technical problems that physics has faced in recent years, and that has created a profound crisis for the profession." This declaration requires some additional background. While Kuhn was familiar with the debates over the interpretation of quantum mechanics, he had been silent on them, while other philosophers of science, such as Paul Feyerabend and Karl Popper had taken stands over them. Thus Kuhn's statement is a pearl from the point of view of the history of philosophy of science; it is a Freudian slip. By Freudian slip, I mean, Kuhn was acting as a physicist committed to the dominant, among the physicists, interpretation of quantum mechanics, and not as a philosopher.[51]

Bohm had not attended the symposium. It fell on the shoulders of Jeffrey Bub to stand up in defense of Bohm's ideas. Bub was successful in the challenge, as reported by Suppe to Bohm: "Your session went quite well ... the discussion of your paper was among the best of any session. Much of the credit is due to Jeffrey Bub, who did a very impressive job of standing in for you." Bub was trained to spot Kuhn's slip. Bub had recently obtained his Ph.D. working on the interpretation of quantum mechanics and was well aware how the physics community was biased against the research on the foundations of quantum mechanics. Bub's answer on this point deserves full transcription[52]:

> Something about what Professor Kuhn just said really disturbed me: it seemed to me that he was suggesting that one should really only care about what most physicists are concerned about today. Now I really find this very puzzling! What significance for the philosophy of science is it that today or in the last thirty years most physicists have been concerned with the particular sorts of problems and that certain problems have been dropped as irrelevant—namely, the problem of the interpretation of the quantum theory [...] I think this is really a very peculiar sort of way to look at the history of physics or the way physics is carried out and then to take what most physicists are doing as some sort of standard on which to base a methodology in the prescriptive sense.

---

[50]Bohm (1974, 378–383) and Kuhn (1962).

[51]Kuhn is cited from Suppe (1974, 409–411).

[52]Bohm to J. Bub, on 19 April 1969, Folder C.137, Bohm Papers. Bub's text is cited from Suppe (1974, 412–413).

Bub's retort was an anticipation of what philosopher and historian of science Mara Beller construed years later. Beller accused Kuhn of building his philosophy of theoretical change in science mirroring it in the current practice of the dominance of the orthodox views on the interpretation of quantum mechanics. According to Beller's words, "the notion of paradigm has not only clear totalitarian implications but also dogmatic ideological roots." Beller founded this strong statement in her identification of the "close historical links [...] between the notion of incommensurable paradigms and the ideology of the Copenhagen dogma."[53]

The symposium Bohm and Ted Bastin organized in mid 1968 in Cambridge may provide a glimpse of Bohm's views on the foundations of quantum mechanics at a moment slightly before he conceived a new approach based on the ideas of wholeness and order. The symposium was held in July 1968, the volume with the papers was published in 1971, and Bohm presented the new ideas in July 1970, in Varenna, Italy. Therefore his papers at the Cambridge symposium are kind of snapshot of his views between July 1968 and early 1970. At the symposium Bohm talked about two subjects. The first was hidden variables, along the lines of his work with Bub, and the second was titled "On Bohr's views concerning the quantum theory," which was a substitute for the joint paper with Schumacher on the failure of communication between Bohr and Einstein.[54] Thus Bohm no longer talked on the idea of discreteness of space-time while Basil Hiley talked on "discreteness, phase space and cohomology." If the published paper does not give a clear hint of the direction Bohm's thoughts would take, his correspondence, particularly with Bub and Schumacher, however, are full of clues. Surely his conversations with Basil Hiley were along the same lines, but we have no written records of these. We know, however, from Hiley's recollections that his own way towards the order approach he would share with Bohm was paved by his own previous experience as a solid-state physicist as he was able to connect his research in statistical mechanics dealing with dimensionality in Ising models with the geometric algebras which would be so instrumental in the order approach. As for Bohm and Bub, they coined the term—DROPS—an acronym for "distinctions, relations, orders, patterns and structures"—to deal with the "relationship between theory and fact." As we will see, order would become the catchword for Bohm's renewed attempt to grasp the quantum theory.[55]

---

[53]Beller (1999, 287–306). On Kuhn and Beller, see Freire Junior (2016).

[54]Bastin (1971). The two subjects are in the papers (Bohm 1971a, d). In the latter paper, Bohm referred, "for a more detailed discussion of this question," to the joint manuscripts D. Bohm and D. Schumacher, "On the role of language forms in experimental and theoretical physics," and "On the failure of communication between Bohr and Einstein." Neither were published. The former is a 27 pp typescript deposited at folder B.88, Bohm Papers, while the latter is a 9 pp typescript, folder B.44, Bohm Papers. Hiley's paper is Hiley (1971).

[55]Bohm to J. Bub, 16 Dec 1966, Bohm Papers, Folder C.130.

5.5 The Quest for a New Approach—Implicate and Explicate Order in Physics

Having abandoned his early causal interpretation, including the modified version with a chaotic subquantum level he had worked on with Vigier, Bohm developed a new approach to the interpretation of quantum mechanics using the idea of putting order as the most basic concept to think about physical theories. The new idea was titled "new order in physics," or, more precisely, the idea of an "implicate and explicate order." Later on, to this title he added the word wholeness and the book presenting it to a wider audience received the title of "Wholeness and the Implicate Order." In this approach, rather than using these theoretical lenses to interpret quantum mechanics, Bohm saw this physical theory as an indication of the centrality of the concept of order and of the need of a new order in physics. In his recollections, Bohm considered this approach to have matured between 1963 and 1965.[56] Indeed, we may see in a talk Bohm gave in 1965 at a conference in Kyoto the conceptual seeds of his later ideas. Bohm presented his talk in a very programmatic manner, combining the idea of discreteness, which will be later abandoned, and that of structural process and order. According to Bohm, "in this paper, a possible new line of development will be sketched, in which it is suggested that these problems can perhaps be solved in terms of the notion of space-time as a discrete structural process." Then the centrality of the idea of order is introduced: "it begins by going axiomatically into the notion of order, which is much deeper than that of measure." Even the idea of totality, later called wholeness, is already suggested, while along the lines of the inseparability between the thought and the universe. According to Bohm, "the tendency to separate these two kinds of structures as if they belonged to different worlds that entered awareness in entirely unrelated ways has, over the past few thousand years, led to enormous confusion in human thinking about the question of 'mind' and 'matter'. Far from being separately existent and structurally unrelated, these are internal and external aspects of one overall creative process, *the totality of all that is*."[57]

However, as far as we were able to see, there were no publications before 1970 presenting the new approach in a systematic manner. In the meantime there possibly were hesitations and dead ends. In fact, in July 1968, Bohm talked at this colloquium he organized with E. Bastin at Cambridge and made no mention of these new ideas. In the Varenna summer school dedicated to the foundations of quantum mechanics, held in July 1970, the new approach was eventually presented to an audience of physicists and physics students interested in the subject of conceptual issues in the foundations of quantum mechanics. This talk was followed by a couple of papers presenting the new ideas in a systematic manner.[58] Unfortunately, we have few annotations, drafts, or letters to help us to analyze Bohm's evolving ideas. However, in his later recollections he came back to this evolution a number of times. Our presentation

---

[56]Bohm (1980). Interview of David Bohm by Maurice Wilkins on 1987 January 30, Session XI, AIP, College Park, MD, USA.

[57]Bohm (1965b).

[58]Bohm (1971a, b, c, 1973).

here is thus strongly dependent on Bohm's later recollections. We also have the recollections of Jefferey Bub, at the time Bohm's doctoral student, which is revealing of the direction of Bohm's thoughts as well as of the manner Bohm groped for the mathematical basis for this approach[59]:

> At the time Bohm was no longer interested in hidden variables. He was trying to develop a general framework for physics based on a discrete space-time structure for events and held a weekly seminar where he discussed ideas on algebraic topology using Hodge's book on harmonic analysis. It was rather too abstract for me. We graduate students tried to make sense of Bohm's ideas with Hiley, but it seemed that every few days ideas he had talked about earlier were scrapped for new ideas, so it was rather frustrating.

Indeed, as late as 1966 Bohm was still thinking of discreteness of space-time as the basis to properly understand the quantum. On October 12, 1966, Bohm wrote to the Italian physicist A. Loinger, as part of the criticisms he and Bub had done on the DLP work, which we have already mentioned: "A generalized field theory in the Einsteinian sense is the probably real alternative to the Copenhagen point of view, though it is my view also that this field should refer to some sort of discrete space time, rather than to a continuum." He added "there was no mathematics that could handle such a theory" and he was "at present working on such mathematics." The quest for a discrete space-time structure, which meant the assumption of a minimum length instead of the continuous space where any length may exist, was not pursued by Bohm alone. This idea had been an open possibility for many physicists who had tried to combine the space-time arena typical of general relativity with the failure of pictures, thus with no or limited reference to space-time, which had been part of standard quantum mechanics. This quest has been charted and discussed, for instance, by the philosopher of science Amit Hagar in his book *Discrete or continuous? The quest for fundamental length in modern physics*. According to Bohm, "maybe quantum mechanics is the physical manifestation of the topology of space-time as gravity is of the metric," and furthermore, in a fundamental quantum mechanical approach, "space-time would be the basis out of which we would abstract quantum mechanics as one of the properties of matter."[60]

Bohm did not succeed in this attempt; according to his recollections, "it seemed it was an interesting clue but some way limited. So it seemed somehow the topological and the measured properties we didn't know how to bring together." The new approach matured first not through a new mathematics but as concepts, strongly dependent on analogies. Its mathematical basis remained an open subject for research, as we will see later. In fact the key concept was that of order and its distinction of implicate and explicate orders. The first insight came from the appreciation of two phenomena, now well-known among Bohm's readers, the experience with ink

---

[59]These recollections were obtained by the author with Jeffrey Bub in 2002 and 2014, see (Freire Junior 2015, 58).

[60]Bohm to Loinger, 12 October 1966, C.130, Bohm Papers. Interview of David Bohm by Maurice Wilkins on 1987 January 30, Session X, Niels Bohr Library & Archives, AIP, College Park, MD, USA. Hagar (2014).

droplets in a rotating and transparent cylinder full with glycerin and the holographic images. Still according to Bohm's recollections,[61]

> I would like to weave together the physical intuition and the mathematics. It seemed that one would try to find some direct interpretation of the mathematics of quantum theory in this way, in terms of space, the properties of space. Then I can remember (but I can't remember the date) I was watching a program on the BBC where they were talking about spin echo, but they gave it an example. This example of the droplet which was immersed in glycerin which spread out and then came back together again. It kind of pulled it into the glycerin and then unfolded. That sort of stuck in my mind that that would be very significant. [...] Then sometime later, a time which I can't remember, I began to think about the hologram. [...] And it struck me that the essential point about the hologram was not its three dimensionality but rather that each part of the hologram contained the image of the whole.

These insights did not come out of the blue. Bohm's evolving ideas were influenced by the writings of Hegel and he also had been influenced by the emphasis the artist Charles Biederman had put on the concepts of order, structure, and creativity, particularly in the arts. In the first public presentation of these new ideas, for instance, Bohm acknowledged to Biederman the suggestion to perceive order as "to give attention to similar differences and different similarities." From both the ink droplet and the hologram phenomena Bohm exploited the idea that the apparent reality—the explicit order—is already contained in a kind of hidden structure—the implicit order. Bohm used implicit, implicate, or enfolded synonymously; and the same way for explicit, explicate, or unfolded. In the case of these two phenomena we have the following analogies: the droplet disappears in the rotating cylinder, thus it is in an implicit order, and the ink can reappear in the form of a droplet, thus assuming an explicit order. With the hologram the connections are not so clear cut but you may say that each part of the image, thus the implicit order, has information about the whole image, the explicit order. Bohm then went further to say that the implicit order was more fundamental than the explicit order. As he recollected, "so I said let's turn it upside down and say there's a fundamental. The basic order is the enfolded order. And the unfolded order emerges from it."[62]

Bohm chose to publish the two seminal papers where he presented his new approach in a recently created journal entitled "Foundations of Physics," a novelty we will comment on later. He began with a historical overview contrasting the ordered world of the Ptolemaic-Aristotelian cosmology with the mathematical space adopted in the inception of the modern mechanics and then set out on a conceptual discussion of order, measure and structure—always appealing to etymological considerations to base his analysis—to understand through these lenses classical physics, relativity, and quantum mechanics. Bohm's strong point is the conceptual incompatibility

---

[61] Interview of David Bohm by Maurice Wilkins on 1987 January 30, Session X, AIP, College Park, MD, USA.

[62] Bohm (1971b, 416). Biederman's ideas are referred to Biederman (1948). However, in the talk at the Kyoto conference the same ideas were introduced—"one may say that order is based on a set of similar differences leading to different similarities"—without further reference (Bohm 1965b, 262). Interview of David Bohm by Maurice Wilkins on 1987 January 30, Session X, AIP, College Park, MD, USA.

between general relativity and quantum field theories, which is because "there is no consistent means of introducing extended structure in relativity, so that particles have to be treated as extensionless points. This has led to infinite results in quantum field-theoretical calculations." At this crucial step in his argument, Bohm suggests we need to go beyond both relativity and quantum mechanics. The first should be reduced to a limiting case: "the limitation to the speed of light will hold on the average and in the long run. Thus, relativistic notions will be relevant in suitable limiting cases." The second, quantum mechanics, should not be considered a sound basis for future theories because the renormalization procedures to overcome the infinities contain "arbitrary features," which are reminiscent of the Ptolemaic epicycles, and the persistence of the controversy around the evolution of the quantum states during the measuring processes. Bohm's bold conclusion is that the development of physics will require giving up "both the basic role of signal and that of quantum state," and "to find a new theory that goes on without these will evidently require radically new notions of order, measure, and structure."[63]

In the follow-up of these ideas Bohm argued that quantum mechanics indicates the need for a new order, "implicate" or "enfolded" order, explaining this idea through the analogy with holograms, and that such an order "should be considered as the independent ground of existence of things, while the ordinary explicate order is what should be considered as dependent." Finally, Bohm explained the kind of mathematics that should be adopted to develop this new order: "the implicate order is expressed naturally in terms of an algebra similar to that of the quantum theory, which is, however, subjected to generalizations going beyond the limits of what has meaning in this theory." Thus is no longer the case of searching for discrete space-time approaches, now the challenge remains in these generalized algebras.[64]

This is not the proper place to analyze Bohm's achievements in this direction, particularly because it was a work-in-progress during the rest of his life, and still is in the hands of Basil Hiley. However, this is the right moment to introduce the role played in Bohm's works by Hiley, his collaborator and a Lecturer at Birkbeck College.

The work with the generalized algebras would take time, around ten years, to bear fruits in terms of publications. Indeed, implicate and explicate order would have remained mere philosophical or scientific intuitions if it had not been for the mathematical elaboration they later received. To accomplish this Bohm did not work alone. He counted on the collaboration of Basil Hiley, who was born in Burma, then part of the British Raj. He came to England when India gained independence and did his degree and doctoral studies at King's College working with the theory of condensed matter, but he was interested in abstract mathematics and foundational physics. He attended a lecture by Bohm at the end of his studies and was spellbound. Professional interaction with Bohm, however, came later, after Hiley was hired by Birkbeck College in 1961. Bohm was also there and they became collaborators. At the beginning of their collaboration there was no connection with Bohm's previous

---

[63]Bohm (1971c, 1973).
[64]Bohm (1973).

work on the causal interpretation. "When I started with Bohm we did not mention or discuss his '52 Hidden Variable approach at all" and "for about the first 10 years we didn't discuss the Hidden Variable Theory hardly at all," Hiley stated. Furthermore, according to Hiley's recollections, he "was brought up in an atmosphere where it was generally agreed that there was something basically wrong with the '52 paper of Bohm." Instead of hidden variable models, Hiley engaged with new mathematical objects with Bohm and the mathematician Roger Penrose, in a seminar they informally ran on Thursday afternoons.[65]

Bohm and Hiley's strategy was to analyze the algebraic structures behind quantum mechanics' mathematical formalism and subsequently look for more general algebras which could be reduced to the quantum algebras as special cases. This was informed by the fact that they did not want to take any kind of space-time geometry as assumptions in their reasoning. Instead they tried to develop algebraic structures from which space-time could emerge. Here the algebraic primary structure would be the implicate order and the emerging space-time geometry would be the explicate order. With the benefit of hindsight, Hiley's unique contribution can be seen in this sense. Indeed Hiley was, and still is, the mathematical mind behind the research program related to the idea of order. A number of different factors also contributed to the development of this mathematical approach, such as new and mathematically talented students including Fabio Frescura, interactions with the mathematician Roger Penrose at Birkbeck College, and inspiration from the Brazilian physicist Mario Schönberg's early works on algebras and geometry. It is noticeable that in parallel, without much connection with Hiley and Bohm, some former students of Schönberg were working along analogous lines. Thus, in one of their works, in 1985, the Brazilian physicists Luciano Videira, Alberto da Rocha Barros and Normando Fernandes made the case, dealing only with relativity, "that instead of departing from a given postulated space-time and then inferring the associated particle dynamics, we should start by imposing a certain physics and then try to determine its related geometry. In other words, geometry should be considered as an aspect of dynamics." They concluded that "this point of view reminds us of Leibniz's conception of dynamics," a conclusion which was appreciated by David Bohm. Highly sophisticated from the mathematical point of view, Bohm and Hiley's approach, however, suffered from little contact with experimental results, which could help to inform the mathematical choices to be made. Still, on this aspect, Bohm and Hiley's approach were not alone in the history of 20th century physics. In fact, around the same time of their work, theoretical physicists have developed a research program, string theories, which present the same characteristic about the relation between theoretical developments and experimental results.[66]

---

[65] See Bohm and Hiley (1981), Frescura and Hiley (1980a, b). Interview of Basil Hiley by Olival Freire on 2008 January 11, Niels Bohr Library & Archives, AIP, College Park, MD USA, www.aip.org/history-programs/niels-bohr-library/oral-histories/33822. Interview of Basil Hiley by Alexei Kojevnikov on 2000 December 5, Niels Bohr Library & Archives, AIP, College Park, MD USA, www.aip.org/history-programs/niels-bohr-library/oral-histories/31624.

[66] See Bohm and Hiley (1981), Frescura and Hiley (1980a, b). Reference to Schönberg is in Frescura and Hiley (1980b). Videira et al. (1985).

5.6   Changing Cultural Landscapes: Bell's Theorem and the Rise of the Quantum Dissidents

While Bohm was beginning a new campaign to enlighten the puzzle of the quanta, the cultural landscape relevant for this research was dramatically changing and Bohm would be pressed to react. On the one hand there was the appearance of Bell's theorem and its experimental tests, everything resulting from Bell's second paper, which Bohm had failed to consider in his Bohm and Bub 1966 paper on measurement and hidden variables. This paper is nowadays considered one of the most basic works in quantum theory since the inception of quantum mechanics and is the base for what the French physicist Alain Aspect would later call "the second quantum revolution." In the early 1970s, the first experiments on Bell's theorem brought conflicting results, which awakened further the debate about its relevance and significance. In addition, an array of new factors—scientific, professional, generational, and political—had heightened interest in the research and debates on the quantum foundations. This was a very different cultural scene from that which Bohm had experienced in early 1950s when he had launched the campaign for the causal interpretation.[67]

5.6.1   Bell's Theorem

In the work where John Bell criticized von Neumann's proof against the hidden variables, Bell continued to ask what features should be required from models with hidden variables, if these models were to be physically interesting. "The hidden variables should surely have some spatial significance and should evolve in time according to prescribed laws." He recognized that "these are prejudices," but added "it is just this possibility of interpolating some (preferably causal) space-time picture, between preparation of and measurements on states, that makes the quest for hidden variables interesting to the unsophisticated." As the ideas of space, time, and causality had not been relevant in von Neumann's assumptions, he attempted to determine what implications follow from hidden variables related to such ideas. Bell had been impressed by Bohm's early causal interpretation since it was published. Later on, he would say: "In 1952 I saw the impossible done," and "Bohm's 1952 papers on quantum mechanics were for me a revelation." Bell thus wrote the wave function of a hidden-variable model for the case of a system with two spin-½ particles. Then, he showed that this wave function is in general not factorable and presents a grossly non-local character, since "in this theory an explicit causal mechanism exists whereby the disposition of one piece of apparatus affects the results obtained with a distant piece." As the state of two spin ½ particles could represent a system similar to that suggested by Einstein, Podolsky, and Rosen, in 1935, Bell concluded, "in fact the

---

[67]This section is largely based on my The Quantum Dissidents (Freire Junior, 2015), particularly on Chaps. 4, 6 and 7. For the "second quantum revolution," see Aspect (2004), on Bell's biography, see Whitaker (2016).

Einstein–Podolsky–Rosen paradox is resolved in the way which Einstein would have liked least." Bell finally asked himself if non-locality is the price to be paid for the existence of hidden-variable theories, such as Bohm's, and admitted that there was no evidence for this.[68]

Bell took then the next logical step: to isolate which reasonable assumption was behind Einstein's argument and check the compatibility between this assumption and quantum mechanics. Following Bell's presentation in his second paper (Bell 1964), which was published before the first paper due to circumstances, he considered the "vital assumption" when dealing with a two-particle system to be what is being measured in one of them does not affect the other. He recalled Einstein's dictum, according to which, "on one supposition we should, in my opinion, absolutely hold fast: the real factual situation of the system S_2 is independent of what is done with the system S_1, which is spatially separated from the former." In the next step, Bell built a simple model of a hidden variable theory obeying such a supposition and showed that its results conflict with quantum mechanical predictions in very special cases. This is Bell's theorem: no local hidden-variable theory can recover all quantum mechanical predictions. In a very rough description, Bell's theorem can be derived when one considers a hidden variable model of a system with two spin-½ particles in the singlet state moving in opposite directions, a system that is analogous to the system suggested in the Einstein–Podolsky–Rosen argument, which Bohm had shown in his 1951 textbook. Bell then wrote a function that is the expectation value of the product of spin components of each particle, and using different spin components derived an inequality with this function. The theorem is demonstrated when one uses quantum mechanical predictions in such inequality because some quantum mechanical predictions violate this inequality. Since then, many other analogous inequalities have been obtained, adopting somewhat different premises; thus today we usually speak of Bell's inequalities as the quantitative measurement of Bell's theorem.[69]

The implications of Bell's theorem failed to be immediately recognized by many. It took a few years for some American physicists, led by John Clauser and Abner Shimony, to fully grasp the fact that experimental results were necessary to settle the dilemma: either quantum mechanics predictions or local hidden variable theories, or still, either quantum mechanics or local realism. Furthermore, four physicists, Michael Horne and Richard Holt in addition to Clauser and Shimony, noted that no available experiments could check Bell's inequalities. To make such a test you would need to take a pair or particles and measure either their spin components or polarization in angles intermediate between 0° and 90°, a task which had not yet been carried out. These four physicists produced a paper, published in 1969 which became a milestone in the history of the research into the foundations of quantum mechanics. In addition to the considerations we have presented here, they adapted Bell's inequalities in a real experiment, which required the introduction of additional assumptions. They recommended that the best experiments should be performed

---

[68]Bell (1966, 1982, 1987).
[69]Bell (1964).

in the field of optics, measuring light polarization in certain angles. Clauser and Holt, at Berkeley and Harvard, respectively, rushed to perform such experiments and by 1972 they had conflicting results: Clauser's supporting quantum predictions and Holt's violating quantum mechanics. These results attracted considerable attention and other experiments were planned. Clauser went on to reproduce Holt's experiment, suspecting the existence of experimental biases; Edward Fry in the US planned an improved experiment using a tunable laser to obtain a cleaner source of pairs of photons; Alain Aspect in France began to plan further sophisticated experiments, particularly following an old recommendation of Bohm and Bell while dealing with EPR-type experiments, namely to rotate the detectors while the photons were in flight; all of these physicists worked with optical photons. Other physicists followed the work of Wu and Sakhnov 1949 experiment and worked with high energy photons. Among others, A. Wilson and colleagues, at Birkbeck College itself where Bohm worked; G. Faraci in Catania; and Madame C. S. Wu, at Columbia University, who was joined by L. R. Kasday to redo the Wu and Sakhnov (1950) experiment, now using high-energy gamma-rays emitted in positron annihilation and using better detectors.[70] Jumping to my conclusions, these experiments ultimately confirmed quantum mechanics predictions. However, in the mid-1970s, with Clauser's and Holt's conflicting results, there seemed no straightforward conclusion that quantum mechanics would endure these tests. In 1974, David Bohm attended a conference in Strasbourg, organized by the French physicist Michel Paty and the Brazilian physicist José Leite Lopes, to mark the 50th anniversary of de Broglie's material waves. There Bohm watched Paty present the following balance of these experiments: "the present balance sheet of the experiments designed to test Bell's inequalities is therefore as follows: three agree with quantum mechanics, and two disagree." Paty followed analyzing, "has quantum mechanics now revealed its limitations, or more exactly, the limits of its field of application?" He conjectured that "This would not be unthinkable a priori, […]. This would also be the case for a theory as powerful as quantum mechanics, which itself is highly powerful, but at the same time probably has a frail basis." Paty, himself, did not support such a conjecture and expressed his trust in quantum mechanics: "However, it may seem doubtful that such an established theory might be questioned in such simple experiments. And in fact quantum mechanics may only appear to be frail; its hold on our conceptions is paradoxically shown in this recent questioning: it is not quantum mechanics which is put into doubt, so much as the basis of these very experiments or at least their interpretation." Bohm would follow along similar lines (Fig. 5.4).[71]

Thinking about the whole situation, which included Bohm's earlier interpretation, which is as nonlocal as standard quantum mechanics is, Bohm and Hiley arrived at the conclusion that nonlocality was a typical feature of quantum mechanics. They concluded this before the announcement of the first results which would unbalance

---

[70]Clauser et al. (1969), Wilson et al. (1976), Faraci et al. (1974), Kasday et al. (1975). On Wu's experiment, see Maia Filho and Silva (2019). For a review of the experiments on Bell's theorem, see Clauser and Shimony (1978) and Freire Junior (2015).

[71]Paty (1977).

Fig. 5.4   John Bell (1928–1990)—On the board, drawing of Aspect's 1982 experiment with two-channel polarizers. *Credits* Courtesy of *Nature*

the results favorably to quantum mechanics. Their paper was submitted on March 29, 1974, and they specifically entitled their paper, "On the Intuitive Understanding of Nonlocality as Implied by Quantum Theory." Due to the relevance of this paper to both the state-of-the-art and the evolution of Bohm's thoughts, I shall make a more detailed presentation of its content. They began by arguing that nonlocality was the utmost quantum signature. They reminded the readers that "it is generally acknowledged that quantum theory has many strikingly novel features, including discreteness of energy and momentum, discrete jumps in quantum processes, wave-particle duality, barrier penetration, etc." But, they warned, "there has been too little emphasis on what is, in our view, the most fundamentally different new feature of all; i.e., the intimate interconnection of different systems that are not in spatial contact." They commented that this was brought to the discussion first with the EPR experiment. It is a conspicuous feature of quantum systems always you take many-body systems and treat them through the 3N-dimensional configuration space required by the Schrödinger equation. They added that interest in this subject had been recently stimulated by Bell's second paper, which Bohm had missed in his earlier papers. Bohm and Hiley argued that Bell had "obtained precise mathematical criteria distinguishing the experimental consequences of this feature of 'quantum interconnectedness of distant systems' from what is to be expected in 'classical type' hidden variable theories, in which each system is supposed to be localizable, i.e., to have basic qualities and properties that are not dependent in an essential way on its interconnections with distant systems." Another major novelty in Bohm and Hiley's paper came when they presented nonlocality as described by Bohm's 1952 causal interpretation. The reason for this was not that this interpretation would be truer than the others, but that it allowed more "imaginative and intuitive concepts" to grasp nonlocality. Otherwise one would be doomed to appeal to "abstract mathematical concepts" to understand what quantum nonlocality is. For Bohm and Hiley,[72]

> In this connection, it might even be useful to consider theories that we do not regard as adequate on general grounds or as definitive in any sense. For such, theories may still give imaginative and intuitive insight into a situation for which there is at present no other way to obtain this kind of insight. It is in such a spirit that we are proceeding in the present paper.

The clever way in which the causal interpretation was being presented is striking; as kind of a heuristic resource to deal with a new and, to say the least, awkward, physical phenomenon, quantum nonlocality, or entanglement, the term once used by Schrödinger and that would later acquire currency. Bohm and Hiley then showed that nonlocality could be understood as a direct consequence of the quantum potential from the 1952 causal interpretation. They showed the quantum potential "does not in general produce a vanishing interaction between two particles" when they are far away from each other and "the quantum potential cannot be expressed as a universally determined function of all the coordinates, [...] Rather, it depends [on the quantum state of the system of all particles] and therefore on the 'quantum state' of the system as a whole."[73]

After this explanation Bohm and Hiley introduced some philosophical and conceptual consequences from their reasoning. First, they argued that quantum nonlocality suggested the adoption of the concept of "unbroken wholeness," which was part of the new approach they had suggested a few years before based on the notions of wholeness and order. Second, they presented the similarities and differences between their view and Bohr's complementarity in a clear-cut manner. "There is a considerable similarity here to Bohr's point of view. Thus, Bohr emphasizes the wholeness of the form of the experimental conditions and the content of the experimental results," recalled Bohm and Hiley. Then they pointed out the differences, "but Bohr also implies that this wholeness is not describable in intuitive and imaginative terms, because (in his view) we must use classical language and classical concepts to describe our actual (large-scale) experience in physics, and these evidently contradict the wholeness that both Bohr and we agree to be necessary." One can note here how far Bohm's reasoning was from the emphasis on the recovery of causality which was typical of the debates over his interpretation in the 1950s. Third, and finally, they were explicit about the still unsolved difficulty with this interpretation, the potential conflict with relativity: "we are quite aware of the limitations and unsatisfactory features of the interpretation." They suggested a way to approach the obstacle: "Can the quantum potential carry a signal? If it can, we will be led to a violation of the principles of

---

[72]Bohm and Hiley (1975). Bell's second paper is Bell (1964).
[73]Bohm and Hiley (1975).

Einstein's theory of relativity, because the instantaneous interaction implied by the quantum potential will lead to the possibility of a signal that is faster than light." Their answer was "one avenue of enquiry is to try to find out to what extent the quantum potential can carry a signal," and they acknowledged that "at present, the answer is, of course, not known."[74]

Their conclusion was a two folded statement. Bohm and Hiley were not attached to the basics of the causal interpretation, as "our attitude is that we can sooner or later drop the notion of the quantum potential (as we can drop the scaffolding when a building is ready) and go on to radically new concepts, which incorporate the wholeness of form–which we feel to be the essential significance of quantum descriptions." This part of the argument may be disputed because one may question the advantages to call the quantum potential to the debate. However, Bohm and Hiley followed to make a statement that is beyond doubt as it encapsulates the conceptual challenge the physicists would face from then on, how to grasp the quantum nonlocality: "We have instead to start from nonlocality as the basic concept, and to obtain locality as a special and limiting case, applicable when there is relative functional independence of the various 'elements' appearing in our descriptions. This means that our notions of space and time will have to change in a fundamental way."[75]

Therefore, thanks to Bell's theorem, Bohm returned to the hidden variables issue. As for his wholeness and implicate order program, hidden variables was a secondary issue, but hidden variables had been brought to the foreground by the appearance of the theorem and its experimental tests, as well as by the interest this whole business had awakened among physicists who were attentive to the foundations of quantum mechanics. Bohm did not balk before the challenge but he would suffer further pressure in reconsidering the hidden variable issue, as we will see in the next chapter.

Before moving to a different issue, let us comment something about the background of Bohm and Hiley's paper. It is noteworthy that Bohm and Hiley only cited the experiment conducted by Freedman and Clauser, which was favorable to quantum mechanics. They did not cite Richard Holt's results, against quantum mechanics, which were not published, but had been reported by Abner Shimony, still as a project, at the Varenna 1970 Summer School, and by Michel Paty, in 1974, when Holt had already obtained the results of his experiment. Thus why were Bohm and Hiley so confidant about the validity of quantum non-locality if the experimental results were still, in 1974, conflicting?[76]

Basil Hiley's recollections may help us to answer this question. Hiley and Bohm "had been discussing quantum non-locality [QNL] since 1964." Indeed, they did not know about Bell's "until a few years later." Their "confidence emanated from the experimental work that Alan Wilson, David Butt and June Lowe were doing at Birkbeck College." Hiley remembers that while the results were published in 1976, they "knew the results long before the paper was published." According to Hiley's

---

[74]Bohm and Hiley (1975).

[75]Bohm and Hiley (1975).

[76]Freedman and Clauser (1972) is cited on Bohm and Hiley (1975, 94). Shimony (1971), Paty (1977).

recollections, "the method they used was to see if the entangled state 'spontaneously localises' as the distance between the two detectors was systematically moved further and further apart. This ensured two things (1) the detection events were space-like separated. This was the first experiment to test this. (2) there could be no possibility of a overlap of the wave function of the individual gamma rays." Still, following Hiley's recollections, "some earlier experiments had shown the entangled state was correct for close, slower detection resolution times; what WBL did was to use faster detectors which were moved sufficiently far apart to ensure space-like separation. They saw no change at all up the 6 m separation."[77]

Hiley's recollections thus brings lessons for the history of science, more precisely for the history of the practice of physics and the complex relationship between theories and experiments. Hiley remarks that "interaction of the theorist and experimenter does not get recorded in final papers, but the regular debates on failures and successes of the unfolding experiment concentrates the mind and clarifies perceptions." This happened with them as they "saw the results unfolding, [they] became more convinced that the feature QNL is there." Furthermore explicitly, Hiley recalls, "notice we were not testing the Bell inequalities—in the planning stages and early preparation we did know about the Bell inequalities. When Bohm and I did discover the Bell paper, Bohm was not very convinced by the Bell argument." Indeed, according to Hiley, "we were testing something much simpler—was there a transition from the entangled state to the local product state, hence our discussion of the notion of spontaneous localisation." Finally, they were paying attention to another possibility of physics which today is almost entirely forgotten in the debates about quantum non-locality, namely that "Hawking was suggesting there could be a many mini-black holes floating around so if we used the Penrose argument that black holes could explain the source of non-unitary transformations, our experiment could provide evidence for or against the existence of mini-blackholes." Hiley finished his recollections pondering, "it was a wild speculation but as the experiment is up and running why not check?"[78]

5.6.2   The Rise of the Quantum Dissidents

In early 1970s an array of different factors led to a surge in interest in the research and debates on the quantum foundations. The singular most influential effect, a scientific one, was the appearance of Bell's theorem and its experimental tests. Other unsolved conceptual issues in foundations were also pressing in the same direction, among them a huge controversy on the quantum measurement problem, which split the orthodoxy concerning the interpretation of quantum mechanics, and the connections between quantum mechanics and gravitation. On this split, Bohm and Bub had been

---

[77]Basil Hiley to Olival Freire, e-mail, 23 March 2019. Wilson et al. (1976). Alan Wilson was killed in a mountaineering accident just after the publication of his paper.

[78]Basil Hiley to Olival Freire, e-mail, 23 March 2019.

actors in the process by suggesting a different approach to the measurement problem. They also had criticized the solution proposed by the Italian physicists Daneri, Loinger, and Prosperi and supported by Rosenfeld. The Italian work had also been criticized by Eugene Wigner. Bohm had a clear perception of these changing times at the opening of the Cambridge conference he had jointly organized with Ted Bastin. Indeed, the physicist Otto Frisch, who had played a major role in the understanding of the nuclear fission, back in 1939, commented: "I understand that at present there exists a controversy, roughly speaking between a group of people which includes Wigner as the best known person and another group centered on Milan in Italy, and that these two have different views on how this reduction [of the quantum state] happens."[79]

The growing interest on foundational issues needed and led to the creation of institutional spaces. The Italian physical society, which promoted the traditional Italian summer schools held in Varenna every year, dedicated one of its 1970 courses to the Foundations of Quantum Mechanics. The French physicist Bernard d'Espagnat was invited to be the director of the school and Bohm was invited to give a lecture there. Basil Hiley attended the course. The American Physical Society opened the pages of its magazine, *Physics Today*, to a debate on the interpretations of quantum mechanics. The debate was opened with an article by Bryce DeWitt, who was by then defending the alternative interpretation of quantum mechanics Hugh Everett had suggested in 1957, and a number of articles and letters ensued DeWitt's paper. The mood of the changing times was captured in one of these letters, published in 1971, written by M. Hammerton, from the Medical Research Center, Cambridge, UK, who wrote,[80]

> The very interesting contributions to the quantum mechanics debate […] exemplify the highly complex and subtle ways in which scientific opinion can change. When I was an undergraduate reading physics 20 years ago, […] the Copenhagen line was "scientific," anything else was meaningless, mumbo-jumbo, or, at best, mistaken. Now the curious thing is that, as far as I am aware, there has been no major finding or theoretical insight that could be held to demolish or supersede this interpretation. Nevertheless, there is now considerable dissatisfaction with it, and a willingness to regard other points of view—for example, hidden variables—as being at least respectable.

Hammerton's complex and subtle ways included the social, cultural, and political unrest around 1970, which forced the community of physicists to make room for subjects so far pushed to one side by physicists, such as foundations of quantum mechanics and relations between science and society. Historians of science have, afterwards, exhibited the workings of some of these factors. The political context which pressed the physics community to create room for the research on foundations, in the case of Italy and the US, has been studied by myself and by Flavio del Santo and colleagues, in the Italian case. David Kaiser has showed how the counterculture movement, influential as it was in the US and in California in particular, was an ambiance conducive to the research on foundations of quantum mechanics.

---

[79]Frisch (1971, 14).
[80]Hammerton (1971).

Hammerton's ways also included the generational change with younger physicists less reverent to the founding fathers of quantum mechanics, as I have argued in my book *The Quantum Dissidents*. New institutional spaces included formal publication vehicles, such as the journal *Foundations of Physics*, launched in 1970. Bohm was on its editorial board and became a regular contributor to the journal. Or, still, informal vehicles, such as the *Epistemological Letters*, animated by Abner Shimony, which became a forum for debates concerning Bell's theorem, but not only this, and its philosophical and scientific implications.[81]

I have coined the term "quantum dissidents" to describe the people who acted, from the 1950s to the 1990s, to push foundations of quantum mechanics from the margins of physics to its mainstream. The dissidents were against the "all foundational issues are solved" approach to the foundations of quantum physics, which had prevailed until then. The dissident physicists thought that these issues were worth pursuing as part of a professional career in physics. By doing so, the latter were questioning the very definition of what good physics was and challenging the established distribution of scientific capital, to use the French sociologist Pierre Bourdieu's notion of scientific fields. The term "quantum dissidents" is a borrowing from the notion of dissidence in politics and religion. They include, among the contemporary of David Bohm, Jean-Pierre Vigier, Hugh Everett, John Bell, John Clauser, Abner Shimony, Heinz Dieter Zeh, Bernard d'Espagnat, Franco Selleri, and Alain Aspect, along with some physicists from the old guard of quantum mechanics, such as Louis de Broglie and Eugene Wigner.[82]

Bohm certainly was, since early 1950 but particularly through the 1960s and 1970s, a notable quantum dissident.

References

Aharonov, Y., Bohm, D.: Time in quantum theory and uncertainty relation for time and energy. Phys. Rev. **122**(5), 1649 (1961a)

Aharonov, Y., Bohm, D.: Further considerations on electromagnetic potentials in quantum theory. Phys. Rev. **123**(4), 1511 (1961b)

Aharonov, Y., Bohm, D.: Remarks on possibility of quantum electrodynamics without potentials. Phys. Rev. **125**(6), 2192 (1962)

Aharonov, Y., Bohm, D.: Further discussion of role of electromagnetic potentials in quantum theory. Phys. Rev. **130**(4), 1625 (1963)

Aharonov, Y., Bohm, D.: Answer to fock concerning time energy indeterminacy relation. Phys. Rev. B **134**(6B), 1417 (1964)

Aspect, A.: Introduction: John Bell and the second quantum revolution. In Bell J.S (ed.) Speakable and Unspeakable in Quantum Mechanics (Revised edition—With an Introduction by Alain Aspect), pp. xvii–xxxix. Cambridge University Press, Cambridge (2004)

---

[81] See *The Quantum Dissidents* (Freire Junior 2015), *How the Hipppies Saved Physics* (Kaiser 2012), and Baracca et al. (2017). The professional context around 1970, concerning the rising interest in foundations of quantum mechanics, is discussed in Freire Junior (2004, 2015).

[82] On the quantum dissidents, see Freire Junior (2015) and Bourdieu (1975).

Baracca, A., Bergia, S., Del Santo, F.: The origins of the research on the foundations of quantum mechanics (and other critical activities) in Italy during the 1970s. Stud. Hist. Philos. Mod. Phys. **57**, 66–79 (2017)

Bastin, T. (ed.): Quantum Theory and Beyond—Essays and Discussions Arising from a Colloquium. Cambridge University Press, Cambridge (1971)

Bell, J.S.: On the Einstein Podolsky Rosen paradox. Physics **1**, 195–200 (1964)

Bell, J.S.: On the problem of hidden variables in quantum mechanics. Rev. Mod. Phys. **38**(3), 447–452 (1966)

Bell, J.S.: On the impossible pilot wave. Found. Phys. **12**(10), 989–999 (1982)

Bell, J.S.: Beables for quantum field theory. In: Hiley, B.J., Peat, F.D. (eds.) Quantum Implications: Essays in Honour of David Bohm, pp. 227–234. Routledge & Kegan, London (1987)

Beller, M.: Quantum Dialogue—The Making of a Revolution. The University of Chicago Press, Chicago (1999)

Besson, V.: L'interprétation causale de la mécanique quantique: biographie d'un programme de recherche minoritaire (1951–1964), Ph.D. Dissertation, Université Claude Bernard Lyon 1 and Universidade Federal da Bahia (2018)

Biederman, C.J.: Art as the Evolution of Visual Knowledge. Red Wing, Minneapolis (1948)

Bohm, D.: A proposed explanation of quantum theory in terms of hidden variables at a sub-quantum-mechanical level. In: Körner, S. (ed.) Observation and Interpretation in the Philosophy of Physics—With Special Reference to Quantum Mechanics, pp. 33–40. Dover, New York (1957)

Bohm, D.: General theory of collective coordinates. In: Dewitt, C., Nozières, P. (eds.) The Many Body Problem [Cours Donnés à l'École d'Été de Physique Thèorique—les Houches—Session 1958], pp. 401–516. Les Houches, Dunod, Paris (1959)

Bohm, D.: Hidden variables in the quantum theory. In: Bates, D.R. (ed.) Quantum Theory—III—Radiation and High Energy Physics, pp. 345–387. Academic, New York (1962)

Bohm, D.: The Special Theory of Relativity, W. A. Benjamin, New York (1965a)

Bohm, D.: Space, time and the quantum theory understood in terms of discrete structural process. In: Proceedings of the International Conference on Elementary Particles, Kyoto, pp. 252–287 (1965b)

Bohm, D.: Creativity. Leonardo **1**(2), 137–149 (1968)

Bohm, D.: On the role of hidden variables in the fundamental structure of physics. In: Bastin, T. (ed.) Quantum Theory and Beyond—Essays and Discussions Arising from a Colloquium, pp. 95–116. Cambridge University Press, Cambridge (1971a)

Bohm, D.: Quantum theory as an indication of a new order in physics. In: d'Espagnat, B. (ed.) Foundations of Quantum Mechanics—Proceedings of the International School of Physics "Enrico Fermi", pp. 412–469. Academic Press, New York (1971b)

Bohm, D.: Quantum theory as an indication of a new order in physics. Part A. The development of new order as shown through the history of physics. Found. Phys. **1**(4), 359–381 (1971c)

Bohm, D.: On Bohr's views concerning the quantum theory. In Bastin, T. (ed.) Quantum Theory and Beyond—Essays and Discussions Arising from a Colloquium, pp. 33–40. Cambridge University Press, Cambridge (1971d)

Bohm, D.: Quantum theory as an indication of a new order in physics. Part B. Implicate and explicate order in physical law. Found. Phys. **3**(2), 139–168 (1973)

Bohm, D.: Science as perception-communication [Plus the debates following the paper]. In: Suppe, F. (ed.) The Structure of Scientific Theories, pp. 374–423. The University of Illinois Press, Illinois (1974)

Bohm, D.: Wholeness and the Implicate Order. Routledge & Kegan Paul, London (1980)

Bohm, D.: Hidden variables and the implicate order. In: Hiley, B., Peat, D. (eds.) Quantum Implications: Essays in Honour of David Bohm, pp. 33–45. Routledge, London (1987)

Bohm, D.: On Creativity. Routledge, London (1996)

Bohm, D., Aharonov, Y.: Discussion of experimental proof for the paradox of Einstein, Rosen, and Podolsky. Phys. Rev. **108**(4), 1070–1076 (1957)

Bohm, D., Biederman, C.J., Pylkkänen, P.: Bohm-Biederman Correspondence. Routledge, London (1999)

Bohm, D., Bub, J.: A proposed solution of measurement problem in quantum mechanics by a hidden variable theory. Rev. Mod. Phys. **38**(3), 453–469 (1966a)

Bohm, D., Bub, J.: A refutation of proof by Jauch and Piron that hidden variables can be excluded in quantum mechanics. Rev. Mod. Phys. **38**(3), 470–475 (1966b)

Bohm, D., Bub, J.: On hidden variables—a reply to comments by Jauch and Piron and by Gudder. Rev. Mod. Phys. **40**(1), 235–236 (1968)

Bohm, D., Carmi, G.: Separation of motions of many-body systems into dynamically independent parts by projection onto equilibrium varieties in phase space. I, Phys. Rev. **133**, A319–331 (1964)

Bohm, D.J., Hiley, B.J.: Intuitive understanding of nonlocality as implied by quantum-theory. Found. Phys. **5**(1), 93–109 (1975)

Bohm, D., Hiley, B.J.: On a quantum algebraic approach to a generalized phase-space. Found. Phys. **11**(3–4), 179–203 (1981)

Bohm, D., Peat, F.: D. Science, Order, and Creativity. Bantam, New York (1987)

Bohm, D., Salt, B.: Collective treatment of liquid helium. Rev. Mod. Phys. **39**, 894–910 (1967)

Bohr, N.: Discussion with Einstein on epistemological problems in atomic physics. In: Schilpp, P.A. (ed.) Albert Einstein—Philosopher-Scientist, pp. 199–242. The Library of the Living Philosophers, Evanston (1949)

Bourdieu, P.: Specificity of scientific field and social conditions of progress of reason. Soc. Sci. Inf. **14**, 19–47 (1975)

Bub, J.: The Daneri-Loinger-Prosperi quantum theory of measurement. Nuovo Cimento B **LVII**(2), 503–520 (1968a)

Bub, J.: Hidden variables and the copenhagen interpretation—A reconciliation. Br. J. Philos. Sci. **19**(3), 185–210 (1968b)

Bub, J.: What is a hidden variable theory of quantum phenomena? Int. J. Theor. Phys. **2**(2), 101–123 (1969)

Bub, J.: Comment on the Daneri-Loinger-Prosperi quantum theory of measurement. In: Bastin, T. (ed.) Quantum Theory and Beyond—Essays and Discussions Arising from a Colloquium, pp. 65–70. Cambridge University Press, Cambridge (1971)

Bub, J.: Interpreting the Quantum World. Cambridge University Press, Cambridge (1997)

Capra, F.: The Tao of Physics: An Exploration of the Parallels Between Modern Physics and Eastern Mysticism. Shambhala, Berkeley (1975)

Carmi, G., Bohm, D.: Separation of motions of many-body systems into dynamically independent parts by projection onto equilibrium varieties in phase space. II, Physical Review **133**, A332–350 (1964)

Clauser, J.F., Horne, M.A., Shimony, A., Holt, R.A.: Proposed experiment to test local hidden variable theories. Phys. Rev. Lett. **23**(15), 880–884 (1969)

Clauser, J.F., Shimony, A.: Bell's theorem—experimental tests and implications. Rep. Prog. Phys. **41**(12), 1881–1927 (1978)

de Broglie, L., Vigier, J.P., Bohm, D., Takabayasi, T.: Rotator model of elementary particles considered as relativistic extended structures in Minkowski space. Phys. Rev. **129**(1), 438–450 (1963)

Duncan, A., Janssen, M.: (Never) Mind your p's and q's: von Neumann versus Jordan on the foundations of quantum theory. Eur. Phys. J. H **38**(2), 175–259 (2013)

Faraci, G., Gutkowski, S., Notarrigo, S., Pennisi, A.R.: Experimental test of EPR paradox. Lettere al Nuovo Cimento **9**(15), 607–611 (1974)

Freire Junior, O.: The historical roots of "foundations of quantum mechanics" as a field of research (1950–1970). Found. Phys. **34**(11), 1741–1760 (2004)

Freire Junior, O.: The Quantum Dissidents—Rebuilding the Foundations of Quantum Mechanics 1950–1990. Springer, Berlin (2015)

Freire Junior, O.: Contemporary science and the history and philosophy of science. In: Blum, A., Gavroglu, K., Joas, C., Renn, J. (eds.) Shifting Paradigms—Thomas S. Kuhn and the History of Science, pp. 105–114, Edition Open Access, Berlin (2016)

Frescura, F.A.M., Hiley, B.J.: The implicate order, algebras, and the spinor. Found. Phys. **10**(1–2), 7–31 (1980a)

Frescura, F.A.M., Hiley, B.J.: The algebraization of quantum-mechanics and the implicate order. Found. Phys. **10**(9–10), 705–722 (1980b)

Frigg, R.: GRW Theory (Ghirardi, Rimini, Weber Model of Quantum Mechanics). In: Dan Greenberger, D., Hentschel, K., Weinert, F. (eds.) Compendium of Quantum Physics—Concepts, Experiments, History and Philosophy, pp. 266–270. Springer, Berlin (2009)

Frisch, O.R.: The conceptual problem of quantum theory from the experimentalist's point of view. In: Bastin, T. (ed.) Quantum Theory and Beyond—Essays and Discussions Arising From a Colloquium, pp. 13–21. Cambridge University Press, London (1971)

Hagar, A.: Discrete or continuous? The Quest for Fundamental Length in Modern Physics. Cambridge University Press, Cambridge (2014)

Hammerton, M.: Still more quantum mechanics. Phys. Today **24**(10), 11–13 (1971)

Herbert, N.: Flash—a superluminal communicator based upon a new kind of quantum measurement. Found. Phys. **12**(12), 1171–1179 (1982)

Hiley, B.J.: A note on discreteness, phase space and cohomology theory. In: Bastin, T. (ed.) Quantum Theory and Beyond—Essays and Discussions Arising from a Colloquium, pp. 181–190. Cambridge University Press, Cambridge (1971)

Hobsbawm, E.J.: On History. Weidenfeld & Nicolson, London (1997)

Jacobsen, A.: Léon Rosenfeld—Physics, Philosophy, and Politics in the Twentieth Century. World Scientific, Singapore (2012)

Jauch, J.M., Piron, C.: Can hidden variables be excluded in quantum mechanics. Helv. Phys. Acta **36**(7), 827–837 (1963)

Kaiser, D.: How the Hippies Saved Physics: Science, Counterculture, and the Quantum Revival. W. W. Norton, New York (2012)

Kasday, L.R., Ullman, J.D., Wu, C.S.: Angular-correlation of Compton-scattered annihilation photons and hidden variables. Nuovo Cimento B **25**(2), 633–661 (1975)

Krishnamurti, J.: The First and Last Freedom [Foreword by Aldous Huxley]. Harper, New York (1954)

Krishnamurti, J.: Freedom from the known, [edited by Mary Lutyens]. Gollancz, London (1969)

Kuhn, T.S.: The structure of scientific revolutions. The University of Chicago Press, Chicago (1962)

Maia Filho, A., Silva, I.: O experimento WS de 1950 e as suas implicações para a segunda revolução da mecânica quântica, Revista Brasileira de Ensino de Física **41**(2), e2018018 (2019)

Margenau, H.: Measurement and quantum states—I & II. Philos. Sci. **30**, 1–16 and 138–157 (1963)

Moody, D.E.: An Uncommon Collaboration—David Bohm and J. Krishnamurti. Alpha Centauri Press, Ojai, CA (2017)

Papaliolios, C.: Experimental test of a hidden-variable quantum theory. Phys. Rev. Lett. **18**(15), 622–625 (1967)

Paty, M.: The recent attempts to verify quantum mechanics. In: Lopes, J.L., Paty, M. (eds.) Quantum Mechanics, a Half Century Later, pp. 261–289. Reidel, Dordrecht (1977)

Peat, F.D.: Infinite Potential: The Life and Times of David Bohm. Addison Wesley, Reading, Massachusetts (1997)

Petersen, A.: Quantum Physics and the Philosophical Tradition. MIT Press, Cambridge, MA (1968)

Schrecker, E.: The Age of McCarthyism—A Brief History with Documents. Bedford/St. Martins's, Boston (2002)

Schumacher, D.L.: Fundamental physics and instrumental technology. Found. Phys. **4**(4), 481–497 (1974)

Schumacher, D.L., Horak, H.G.: Direction of time and equivalence of expanding and contracting world models. Astron. J. **67**(9), 586 (1962)

Shimony, A.: Experimental test of local hidden-variable theories. In d'Espagnat, B. (ed.) Foundations of Quantum Mechanics—Proceedings of the International School of Physics "Enrico Fermi", pp. 182–194. Academic Press, New York (1971)

Siegel, A., Wiener, N.: Theory of measurement in differential-space quantum theory. Phys. Rev. **101**(1), 429–432 (1956)

Suppe, F. (ed).: The Structure of Scientific Theories. The University of Illinois Press, Illinois (1974)

Talbot, C.: David Bohm: Causality and Chance, Letters to Three Women. Springer, Berlin (2017)

Tutsch, J.: Collapse time for the Bohm-Bub hidden variable theory. Rev. Mod. Phys. **40**(1), 232–234 (1968)

Videira, A.L.L., Rocha Barros, A.L., Fernandes, N.C.: Geometry as an aspect of dynamics. Found. Phys. **15**(12), 1247–1262 (1985)

Von Neumann, J.: Mathematical Foundations of Quantum Mechanics. Princeton University Press, Princeton, NJ (1955)

Whitaker, A.: John Stewart Bell and Twentieth-Century Physics: Vision and Integrity. Oxford University Press, New York (2016)

Wiener, N., Siegel, A.: A new form for the statistical postulate of quantum mechanics. Phys. Rev. **91**(6), 1551–1560 (1953)

Wiener, N., Siegel, A.: The differential-space theory of quantum systems. Il Nuovo Cimento **2**(4 Suppl), 982–1003 (1955)

Wilson, A.R., Lowe, J., Butt, D.K.: Measurement of relative planes of polarization of annihilation quanta as a function of separation distance. J. Phys. G: Nucl. Part. Phys. **2**(9), 613–624 (1976)

Wu, C.S., Shaknov, I.: The angular correlation of scattered annihilation radiation. Phys. Rev. **77**(1), 136 (1950)

Chapter 6
The Quest for Compatibility Between the Causal Interpretation and the Wholeness Approach (1979–1992)

In the 1980s, Bohm would spend more and more time dealing with a pressing challenge. The persistent quest to understand the quantum theory would lead him to look for a compatibility between his old causal interpretation and the recent wholeness approach. The pressure rose when some of his and Hiley's students began to produce computerized graphs of the paths derived from the causal interpretation as well as diagrams of its quantum potential. In addition, the continuity of experiments related to Bell's theorem, particularly those by the French physicist Alain Aspect, and the attention they aroused, brought to the foreground two distinct consequences. First, experimental results were confirming the quantum nonlocality, thus discarding the possible local realist theories and demanding the understanding of nonlocality. Second, these same results were also telling that the type of hidden variables used by Bohm in his causal interpretation were not indeed incompatible with standard quantum mechanics. Therefore, it was natural to ask what the problem was, if any, with those hidden variables. Over the previous two decades Bohm had sustained that the quantum potential idea associated with these hidden variables had been a useful notion to show that the usual interpretation was not the only possibility but that this quantum potential was rather artificial. Now, this consideration was no longer enough. Bohm was sensitive to these questions while he was pushing forward his new approach with the publication of the book *Wholeness and the Implicate Order*, in 1980. Furthermore, the increasing interest in quantum mechanics foundations brought to the debate old and new ideas, such as the relative states interpretation suggested by Hugh Everett in 1957 and the spontaneous collapse theory, which meant to modify Schrödinger equation, suggested by Giancarlo Ghirardi and collaborators (the GRW approach). These two proposals attempted to solve the quantum measurement problem and rivalled both the usual interpretation of quantum mechanics and Bohm's proposals. All these intellectual novelties were demanding new responses from Bohm and he reacted accordingly. The result of Bohm's attempts to make his two approaches compatible, the causal and the wholeness interpretations, led him to create a new concept, active information, related to the quantum potential.

© Springer Nature Switzerland AG 2019
O. Freire Junior, *David Bohm*, Springer Biographies,
https://doi.org/10.1007/978-3-030-22715-9_6

Bohm's attempts were consolidated in a major book, written with Basil Hiley, *The undivided universe—An ontological interpretation of quantum theory*, published in 1993. Bohm did not survive to follow the reception of this work. Indeed, the book was practically finished when a heart attack ended his life on October 27, 1992.

His last decade was also a time of a growing recognition of his works among fellow physicists. These years were also the times when Bohm became a public figure revered in circles wider than physics itself. This was in part due to a number of talks he gave and books on philosophy of science he wrote. In the twilight of life, he recovered his American citizenship in court, however, it was too late for him to return to the US.

In the first three sections of this chapter I present the new intellectual landscape around Bohm's quest for the understanding of the quantum. We successively consider the work of his and Hiley's students with computerized graphs derived from the causal interpretation, the publication of *Wholeness and Implicate Order*, and the ongoing series of experiments of Bell's theorem as well as the interest in alternative approaches to quantum mechanics. In Sect. 6.4, I comment on the recognition of his works among the physicists and in Sect. 6.5, I analyze Bohm's interactions with wider intellectual circles, particularly through the diffusion of his philosophical views. Section 6.6 is dedicated to the outcome of his long battle to recover his US citizenship. In Sect. 6.7 I come back to his work on the interpretation of quantum mechanics presenting the ideas expressed in book *The undivided universe* and their reception. Section 6.8 of this chapter focuses on Bohm's failing health.

6.1   The Chrises and the Revival of the Quantum Potential and Its Paths

In the late 1970s Bohm and Hiley were surprised by the independence of mind of two of his students, Chris Philippidis and Chris Dewdney, the Chrises as Hiley called them. The Chrises stumbled upon Bohm's 1952 causal interpretation and wanted to know why this work was not part of the discussions in the research team meetings. At that time, Bohm's attention was devoted to the conceptual and mathematical aspects of the ideas of order and wholeness. The ideas engaged in these reflections varied from discreteness of space-time, to topology, to geometric algebras, but nothing of this had any connection with the early causal interpretation. The Chrises' question however was in the air. After the appearance of Bell's theorem and its experiments, it was reasonable to ask for an interpretation which ultimately had no experimental conflict with the usual interpretation. As basic as this question is, it was up to the younger, junior researchers, and not to their seniors, to ask it. Almost 30 years later this situation was vividly recollected by Basil Hiley[1]:

---

[1]Interview of Basil Hiley by Olival Freire on 2008 January 11, Niels Bohr Library & Archives, American Institute of Physics, College Park, MD USA, www.aip.org/history-programs/niels-bohr-library/oral-histories/33822.

We had a couple of research students working for us, Chris Dewdney and Chris Philippidis. They came to me one day with Bohm's 1952 paper in their hand. And, they said, "Why don't you and David Bohm talk about this stuff?" And I then started saying, "Oh, because it's all wrong." And then they started asking me some questions about it and I had to admit that I had not read the paper properly. Actually I had not read the paper at all apart from the introduction! And when I took it and, so, you know, I was now faced with embarrassment that our research students [Laugh] were putting me in, in a difficult position, and so I went back home and I spent the weekend working through it. As I read it, I thought, "What on earth is wrong with this? It seems perfectly all right. Whether that's the way nature behaves is another matter." But as far as the logic, the mathematics, and the arguments were concerned, it was sound. I went back again to see the two again, I said, "Okay, let's now work out what the trajectories are, work out what the quantum potential looks like in various situations.

The students then surprised Hiley a second time. In fact, Dewdney had discovered Bohm;'s original causal interpretation not directly in the journal *Physical Review*. He had by chance found the book *A survey of hidden-variables theories*, by Frederik Belinfante, on a library table. This was an exceptional book in the sense it was a deep and fair analysis of the hidden-variables issue in quantum mechanics. Belinfante had considered the possibility of calculating the trajectories implied in Bohm's interpretation and considered that "in some simple cases the equations of motion can be integrated. However, he also realized that "in most cases, however, the term $-\nabla U$ in [force equation] causes a motion which is wildly complicated." In fact, in the appendices there is a calculation of Bohm's trajectories but only for the simple case of the Gaussian wave packet. Then the students, Dewdney and Philippidis, went on to calculate the trajectories for the "wildly complicated" case of the most iconic experiment in quantum mechanics, the double-slit experiment. They used the available computer facilities, at that time with a huge pile of punch-cards, and calculated the graphs of the trajectories, thus creating images of quantum phenomena.[2]

This work was immediately well received, particularly among the two teams who had persevered in working with the causal interpretation. They were the French team led by Jean-Pierre Vigier at the Institute Henri Poincaré, in Paris, and the Italian physicists led by Franco Selleri, who was at the University of Bari, in Italy. Curiously, according Dewdney's recollections, while Bohm was not at all surprised with the results—"Yeah, that's how it works"—Vigier and Selleri reacted enthusiastically. Chris Dewdney recalls how, from one moment to the other, he began to receive invitations to give talks presenting their results. In addition to the still pictures, Dewdney produced a short movie illustrating, in a two-slit experiment, how electrons followed a determined path, crossing only one slit each, and nevertheless the accumulation of these paths led to the appearance of an interference diagram, which is the signature of the wave phenomena. These pictures became iconic among every author who wants to present the hidden variable approach to quantum mechanics. Till now, the paper has 170 citations smoothly distributed over time.[3] This ongoing interest in this

---

[2]Interview of Chris Dewdney by Olival Freire on 2016 February 18, Niels Bohr Library & Archives, American Institute of Physics, College Park, MD USA. Belinfante (1973, 91); the calculation for the Gaussian wave packet is in the Appendix G (Belinfante 1973, 194–197).

[3]Interview of Chris Dewdney by Olival Freire, idem. The paper with the graphs is Philippidis et al. (1979). The source for the citations is Web of science, accessed on 21 Nov 2018.

paper is also related to the studies of quantum chaos using Bohm's hidden variables approach (Figs. 6.1, 6.2 and 6.3).

These images brought intelligibility to the quantum phenomena even though they were not derived from new equations. Indeed they were directly derived from the approach and the physical models built by Bohm in early 1950. The full episode reminds us how influential images can be in the practice of science, even in a very mathematical and abstract domain as that of the quantum theory. It also reminds us of the study by the philosopher and historian Peter Galison, who presented the history of the experimental discovery of new subatomic particles throughout the 20th century as an ongoing conflict between two traditions of research, the image and the logic traditions. The former valued the production of pictures highly, even valuing the role of a single good snapshot from a cloud chamber device or from an emulsion plate, the "golden" event, while the latter looked for large numbers to get good statistics, fully disinterested in obtaining images from these microscopic phenomena. Many other episodes in the history of physics would remind us of the power of

Fig. 6.1 Particle trajectories, calculated from Bohm's 1952 hidden variables interpretation, through two slits. These graphs were first obtained by Chris Philippidis and Chris Dewdney, doctoral students of Basil Hiley. *Credits* Philippidis et al. (1979)

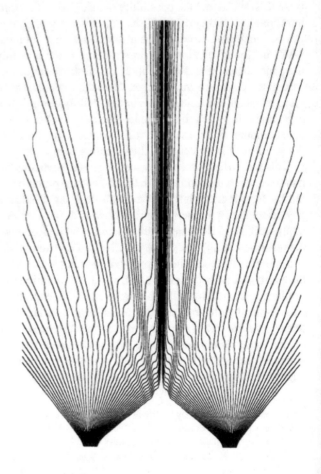

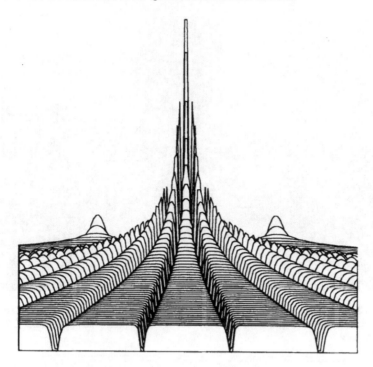

Fig. 6.2 Graphs of the quantum potential, calculated from Bohm's 1952 hidden variables interpretation, of two slits. These graphs were first obtained by Chris Philippidis and Chris Dewdney, doctoral students of Basil Hiley. *Credits* Philippidis et al. (1979)

images, as was the case with Faraday's lines of force and Feynman's diagrams.[4] If we consider the role played by images in the production of science, it comes as no surprise that Bohm and fellow hidden variables supporters would from now on present their approach as bringing an intelligibility which was lacking to the standard presentation of quantum mechanics. This point was made in spite of the fact that in real experiments concerning Bell's theorem, in general dealing with pair of entangled photons, no image is produced. Indeed, as I have argued elsewhere this fits better in the logic tradition, to keep with Galison's terms.[5]

Dewdney's engagement with the causal interpretation and Bohm's works also connect with Bohm's stances in science and society. Dewdney was engaged in the political movements of the times, particularly the collective "Science for People," and was attracted to work on Bohm's ideas also because he saw Bohm as somebody who challenged orthodoxy in science. Dewdney was not alone; we showed in *Quantum Dissidents* that similar relationships among the political and cultural protests of the late 1960s and the rising interest in the research on foundations of quantum mechanics also took place in the Italian and American contexts. It is a kind of a closure of the

---

[4]On the history of the diffusion of Feynman's diagrams, see Kaiser (2005).
[5]Galison (1997) and Freire Junior (2015, 266).

Fig. 6.3 Chris Dewdney, in the late 1970s, one of Basil Hiley's students who rediscovered Bohm's 1952 paper and produced graphs of the trajectories and quantum potential. *Credits* Courtesy of Chris Dewdney

circle, a younger generation being attracted to Bohm's ideas also as a result of the causes which had been so dear to Bohm in the 1940s and 1950s.[6]

6.2   The Book Wholeness and the Implicate Order and Its Reception

In early 1980s David Bohm gathered a collection of his essays in a single volume entitled *Wholeness and the Implicate Order*. The book presented the state of the art of his thoughts on the interpretation of quantum mechanics and his philosophical reflections. The book also reveals the influence of Jiddu Krishnamurti's thoughts on Bohm's reflections. While the book was well received by the wider laymen audience, which can be seen by the number of editions and translations the book has received, Bohm's fellow physicists were colder in their consideration of its content.[7]

---

[6]Interview of Chris Dewdney by Olival Freire on 2016 February 18, AIP, College Park, MD USA. On the British context related to the publication "Science for People," see Werskey (2007). On the Italian and American cases, see Freire Junior (2015), particularly chapter 6: "From the Streets into Academia": Political Activism and the Reconfiguration of Physics Around 1970. On the role of contexts easing the flourishing of the research on the foundations of quantum mechanics, see also Kaiser (2012) and Baracca et al. (2017).

[7]Bohm (1980).

Bohm began his text providing a wide overview of the ideas of fragmentation, or specialization, and of totality, or holism. Here the influence of Krishnamurti on his thoughts on science and society was acknowledged. After discussing the importance of overcoming fragmentation, Bohm concluded the chapter recalling that "As Krishnamurti has brought out with great force and clarity, this requires that man gives his full creative energies. This requires, however, that man gives his full creative energies to the inquiry into the whole field of measure." For Bohm, the value of this consideration was not limited to science. Indeed, still citing the Indian thinker, Bohm concluded, "to do this may perhaps be extremely difficult and arduous, but since everything turns on this, it is surely worthy of the serious attention and utmost consideration of each of us."[8]

The influence of Krishnamurti's ideas on Bohm's physics, if any, may be spotted here in the value of totality, holism, or wholeness. In fact, from the first contact of Bohm with Krishnamurti's writings the inseparability of the observer and the observed was a central issue. It is enough to recall it was this issue which called the attention of Bohm's wife for the Krishnamurti's book at the Bristol library. However, in the emergence of the centrality of the idea of wholeness in Bohm's thoughts several layers of influence overlapped making it difficult to disentangle them to assess the influence of each one. In fact, one may see the inception of the idea of wholeness in the 1952 idea of quantum potential and its action at a distance, thus replicating quantum mechanics predictions for the EPR experiment. The idea of wholeness was key in Niels Bohr's thoughts; precisely the phenomenon which was the object of quantum theory was, according to Niels Bohr, the wholeness of the system under observation and the conditions required for such observation. Still, in the 1950s, with Donald Schumacher and Jeffrey Bub, as we have already seen, Bohm gained a greater understanding of Bohr's thoughts. In this understanding the key teaching to be preserved from complementarity was the Bohrian concept of wholeness. Finally, when Bell's theorem appeared non-locality was brought to the foreground of quantum theory. Thus, we may conclude that the engagement with Krishnamurti's thoughts was an additional influence in Bohm's formulation of the idea of wholeness.

Bohm's reference to Jiddu Krishnamurti in this book requires further clarification. In fact, in the late 1970s Bohm had begun to express, privately, frustration with his interaction with the Indian teacher, particularly with the dogmatic manner in which Krishnamurti considered himself as always being the bearer of the truth. This process would lead, a few years later, to complete estrangement between them. One of Bohm's concerns was that he felt Krishnamurti had "dismissed and tried to push aside my own work in science and philosophy in a way that is not justified, because I think it makes a significant contribution to what is being attempted by all of us." Was Bohm trying to show in *Wholeness and the Implicate Order* with this reference to Krishnamurti that his own work in science and philosophy and Krishnamurti's teachings were in a kind of symbiosis, that is, mutually beneficial? Supporting this conjecture is the fact that the reference to Krishnamurti was not present in the first printing of the book,

---

[8]Krishnamurti citation is in (Bohm 1980, p. 26) and it comes from the book Freedom from Known Krishnamurti (1969).

it was added, without explanation, afterwards. Indeed, in the British printing Bohm wrote in the final paragraph of Chap. 1: "This requires, however, that man gives his full creative energies to the inquiry into the whole field of measure." In the American printing, this sentence was changed to: "As Krishnamurti has brought out with great force and clarity, this requires that man gives his full creative energies to the inquiry into the whole field of measure." However, we do not have enough documentary evidence to go further on this to settle the issue.[9]

It should be noticed that Bohm framed his criticisms towards fragmentation and his defense of wholeness in terms of "Western and Eastern forms of insight into wholeness." While this conceptual framework was widely used at the time, we will see Abner Shimony reproducing it in his review of this book and Bohm using it again a few years later, it is indeed misleading. The question is that such a grand distinction brings more problems than it solves. For instance, it homogenizes Greek philosophy into the Western label (despite Bohm having acknowledged differences between the atomists' and Aristotelian thought), it denies the exchanges between classic Greek thought and cultural influences from Middle East and the north of Africa, it leaves no room for exchanges between Europe, China and India, and it encompasses Chinese and Indian philosophies as the same thing. At the time Bohm used this conceptual framework it had already been criticized as a manner of eurocentrism.[10]

As for the interpretation of quantum mechanics Bohm republished, without changes, both the paper written in the early 1960s presenting the hidden variables interpretation and Bohm's distance from it and the papers from the early 1970s presenting the wholeness and implicate order approach.[11] Thus the volume provided the readers with three papers covering Bohm's evolution concerning the interpretation of quantum mechanics in one volume.

For our later analysis of the synthesis Bohm sought between his hidden variable interpretation and the wholeness approach, it is useful to pinpoint how he presented the concept of the quantum potential, which is a key concept in his early hidden variables approach. Bohm conceded that "it must be admitted that the notion of the 'quantum potential' is not entirely a satisfactory one, for not only is the proposed form, $U = -\left(\frac{\hbar^2}{2m}\right)\left(\frac{\nabla^2 R}{R}\right)$, rather strange and arbitrary but also (unlike other fields such as the electromagnetic) it has no visible source." After stating that "this criticism by no means invalidates the theory as a logical self-consistent structure but only

---

[9]The story of Bohm and Krishnamurti's estrangement is narrated in an "Afterword" David Peat added to his Bohm's biography after he knew of the correspondence between Bohm and the physicist Fritz Wilhelm, where the estrangement is evident. The citation is from a letter from Bohm to Wilhelm, on 28 Jan 1980, which is transcribed in Peat's book. See Peat (1997, 323–330). The British printing of Bohm (1980) has the ISBN number 0-7100-0366-8. All the later printings kept the change with reference to Krishnamurti. Bohm did not add any explanation for the change.

[10]In *Science, Order & Creativity* Bohm dedicated a section to discuss "The Responses of East and West to the Conditioning of Consciousness;" see (Bohm and Peat 1987, 255–260). On the criticism of the oversimplified representation of non-European cultures, see, for instance, Said (1978). I am thankful to Chris Talbot for discussion on this aspect.

[11]The paper on hidden variables is Bohm (1962) and the papers on wholeness and implicate order are Bohm (1971, 1973). These papers are Chaps. 4, 5, and 6, respectively, in Bohm (1980).

attacks its plausibility," Bohm was emphatic: "we evidently cannot be satisfied with accepting such a potential in a definite theory." Thus by 1980 Bohm was not yet embracing the idea he would present later of conceptually interpreting this potential as an "active information."[12]

In another chapter Bohm presented what he called rheomode, a proposal to change the structure of natural languages in order to emphasize the role of actions, thus verbs, instead of entities, namely nouns. For insiders in the debates over the interpretation of quantum mechanics, such a proposal was a clear attempt to overcome Niels Bohr's insistence on the inevitability of classical concepts. For Bohr, these concepts should not be discarded, they should be used subordinate to the complementarity principle. Bohr's point was that they were expressed in natural language and the use of this language was required in order to assure objectivity in science through the communication of experimental results. Throughout the 1960s, with the debates with Jeffrey Bub and Donald Schumacher about Bohr's thoughts, Bohm had never conceded this point to Bohr. The rheomode was thus Bohm's answer to it.

Wholeness and Implicate Order was therefore the presentation of Bohm's current scientific and philosophical ideas for a laymen audience which proved to be quite wide. The same cannot be said of the reception of the book and the ideas of order and wholeness among his fellow physicists. Here, unlike the criticism Bohm had received, for instance, for his early hidden variable interpretation, in the 1950s, his ideas were now received with a certain indifference. Bohm was keenly aware of this, reporting in 1986, "at the time, I remember I gave talks on the implicate order to all sorts of people, like artists and architects and so on, and they appreciated it. See, that all helped. They could all get interested in it. The physicists couldn't because they said, 'It has nothing to do with physics.'" And yet, after the discouragement of the 1950s, Bohm was far more resilient and this was because physics was now only part of his intellectual interests as his dialogues with Krishnamurti occupied another important part. According to Bohm, "well, one point was that it was somewhat discouraging, but it didn't affect me very much because my interest had gone so strongly toward Krishnamurti that physics was not the only point of my life at the time, not that central, so I could say, 'I'll keep on working on it,' because I had other interests."[13]

Apparently, Bohm was not sensitive to the fact that even among most researchers engaged in the research on the foundations of physics, by then a larger and more active community than it was in the 1950s, the wholeness and implicate order was received with caution. The case of Abner Shimony, the physicist who jointly with John Clauser led Bell's theorem to the lab benches, is a good illustration of this reaction. This is particularly because Shimony had also been trained in philosophy and was highly sensitive to the philosophical implications of the research on the foundations of quantum mechanics. Shimony began his extensive review of the book for *Nature* identifying the dominant intellectual trend in the book: "This book is

---

[12]Bohm (1980, 80).

[13]Bohm's citations are in Interview of David Bohm by Maurice Wilkins on 1987 April 3, Niels Bohr Library & Archives, American Institute of Physics, College Park, MD USA, www.aip.org/history-programs/niels-bohr-library/oral-histories/32977-11.

David Bohm's attempt to effect a meeting of East and West. In the basic principles of his metaphysics and epistemology the East predominates. The world is an undivided whole, and any analysis of it into parts (things, atoms, sensations) is an erroneous fragmentation." However, Shimony did not criticize this. The weakness of the book, for Shimony, lay elsewhere, in the underdeveloped character from a scientific point of view of the ideas of wholeness and order. Shimony expressed his trust in Bohm's approach, stating: "Bohm's programme is ambitious, and in the long run it must be assessed by concrete achievements." However, this vote of confidence was guarded because "The sketch which Bohm provides of the mathematical apparatus of his programme [...], is, unfortunately, so cryptic that I can make no judgement about its prospects. I can only say that he has achieved enough as a physicist to deserve both time for pursuing his programme and attention for his claims of concrete results." Another review along the same lines came from the philosopher of science Martin Curd, from Purdue University. Curd was optimistic that the new order approach "might lead to the development of theories that go beyond quantum mechanics and relativity and that, from this new viewpoint, might effect a reconciliation between them at some deeper level of reality." However, he considered the suggestions in this direction presented in the book to be too vague. According to him, "Bohm's contribution to this project is, by his own admission, at best and tentative. His ideas are often expressed nonmathematically through the use of metaphors and partial analogies [...] and are frequently characterized by considerable imprecision and fuzziness." Despite these caveats, the reviewer conceded that "it would be churlish to condemn on grounds of vagueness and imprecision a work whose primary aim is to stimulate inquiry."[14]

Among the audience beyond physicists, the book was, as we have already noted, better acclaimed. To illustrate this, let us cite two cases. Steven N. Rosen, from the Psychology, Sociology and Anthropology Department at the College of Staten Island, City University of New York, received the book with wider considerations: "we live in an era of uncertainty and alienation, a time when all of our standards have been opened to question our most hallowed institutions and cherished beliefs." He described Bohm as somebody who was "striving to bring into harmony the contemporary knowledge of our post-Einsteinian world and the perennial wisdom of the ages." In an extended essay, Rosen followed connecting both Bohm's philosophical considerations about fragmentation and wholeness and his algebraic incursions with the psychological ideas of Carl Gustav Jung. Bohm's book was also very well received by John Seekamp, president of the Swedenborg Foundation, who invited him "as an honored guest and participant to our Symposium on 'Science and Spirituality: A Search for Unity.'" According to Seekamp, "the Symposium will honor the tricentennial of the Swedish scientist and mystic Emanuel Swedenborg." Still, this symposium would be unique "to bring together the major scholars in the fields

---

[14]On Shimony's intellectual wide interests, see Shimony (1981, 1993) and Curd (1981). For a more critical review, see Park (1981).

of the new physics, the new biology, the new psychology, philosophy, metaphysics, life process, and near-death experience face-to-face in one seminar."[15]

6.3   Research on the Foundations of Quantum Mechanics Reaches a New Stage

Throughout the 1980s the research on the foundations of quantum mechanics reached a new stage awakening a wider interest among the physicists. By the same token, it challenged Bohm to resume his approaches in order to make them dialogue with the new developments. These novelties may be broadly grouped into the following subjects: Alain Aspect's 1982 experiments on Bell's theorem; ongoing interest, from early 1970s on, in Everett's interpretation; the increasing interest in the use of quantum mechanics in cosmology and attempts to unify quantum theory and gravity; the appearance of concrete models modifying the Schrödinger equation to deal with the measurement problem, particularly the proposal led by Giancarlo Girardi; the maturing of research into the decoherence effect; and the early works relating foundations of quantum mechanics to information. It was thus a very different intellectual landscape from that of the 1950s when Bohm had suggested his causal interpretation which had been poorly received by those among the physics community. All of these new developments and results Bohm had to react to. However, first let us present an overview of these before analyzing Bohm's reactions.

Alain Aspect's 1982 experiments on Bell's theorem contrasting quantum mechanics predictions and the class of local hidden variable theories had been long expected among the physicists who followed the experimental status of Bell's theorem.[16] He had presented the project of these experiments in 1976 at the Erice Thinkshop on Physics, which was organized by John Bell. Aspect promised to perform the kind of experiment Bell was looking for, namely to change the polarization filter directions while the pair of photons were in flight from their common source to the detectors. These changes should happen in time intervals less than the time light would take from the source to the detectors, thus preventing possible unknown interactions between sources and detectors which could bias the experiment and influence the violation of Bell's inequalities obtained so far. In his hometown, Paris, there was a keen interest in the results as Aspect had convinced the influential French physicist Claude Cohen-Tannoudji of the relevance of his experiments. Thus, in June 1980, when the first results were about to appear, a colloquium was held at the prestigious Collège de France to discuss the conceptual implications of quantum physics and

---

[15]Rosen (1982). John R. Seekamp to Bohm, 16 Oct 1987. Both documents are at Bohm Papers, Folders A.73 and B.79, respectively, Birkbeck College, London. Seekamp sent attached to his letter proofs of a review of Bohm's book, which was scheduled to appear in Chrysalis, the foundation's journal, in 1988.

[16]For a detailed description of the early experiments dealing with Bell's theorem, see Freire Junior (2015, 235–286).

particularly Bell's theorem and Aspect's experiments. Eventually Aspect came with three different experiments. The first was a slighter changed replication of previous experiments with better accuracy, the second had a hugely improved design detecting photons on two different channels on each side of the source of the photons, and the third was with the time-varying analyzers. All three experiments violated Bell's inequalities and confirmed quantum mechanics predictions.

These results brought sudden and wide recognition both to the subject, the confirmation of the weird entanglement as a quantum effect, and to the physicist who worked on these breakthroughs. As for Aspect, the Eighth International Conference on Atomic Physics, held in Sweden, in 1982, chose him to be one of the invited speakers to report his experiments and the subject was a prominent part of the conference final summary presented by Arthur Schawlow, who had been awarded the Nobel Prize the year before. Years later, the prestigious Wolf Prize acknowledged three different generations of experimental physicists who contributed to establishing quantum entanglement as a physical property relevant for basic research as well as for technological applications in the new domain of quantum information. The first was John Clauser, who performed experiments in early 1970s, the second was Alain Aspect, and the third one was Anton Zeilinger, who performed new experiments from early the 1990s on, thus beyond the scope of our story.

The wide acknowledgment of quantum entanglement as a physical effect, or, put otherwise, the admission that no local hidden variables theory could replicate all quantum predictions, a statement which was enhanced by Aspect's experiments, had an important side effect. Bohm's 1952 hidden variable theory could no longer be seen as having potential conflicts with quantum mechanics; indeed, it was as nonlocal as standard quantum mechanics, something Bohm and Hiley had remarked the decade before, in 1975. After the announcement of the results of Aspect's experiments it was the moment for John Bell do the same. He took the opportunity of a special issue of the journal *Foundations of Physics* dedicated to the 90th anniversary of Louis de Broglie to publish a paper with the meaningful title "On the impossible pilot wave." In this paper, Bell revealed how much he had been influenced by the appearance of Bohm's 1952 work, reviewed the theoretical and experimental works related to the possibility of introduction of hidden variables, and boldly concluded: "Of the various impossibility proofs, only those concerned with local causality seem now to retain some significance outside special formalisms. The de Broglie-Bohm theory is not a counter example in this case. Indeed it was the explicit representation of quantum nonlocality in that picture which started a new wave of investigation in this area." But the remaining issue was, why it was so? Or, still, how could one interpret the quantum potential, which was ultimately responsible for non-locality in Bohm's early hidden variable theory? This was the kind of question Bohm needed to deal with now.[17]

In the mid 1980s Bohm's alternative interpretations of quantum mechanics, either the hidden variables or the wholeness, were not the only ones available in the physics market. One of the rival approaches had been suggested by Hugh Everett in the late

---

[17]Bohm and Hiley (1975) and Bell (1982).

1950s in his doctoral dissertation at Princeton. He dispensed with the presumption of a sudden, indeed instantaneous, evolution of quantum states during measurements, a presumption von Neumann's had dealt with in the creation of a new operator, the projection operator. For Everett all the evolutions of quantum states should be described by the unitary evolutions ruled by Schrödinger equation. The cost of this was multiplying levels of reality, which Everett did not consider a problem as this multiplicity was not observable due to the very mathematical formalism of quantum mechanics, a situation Everett's named the "relative states" interpretation. During its elaboration, Everett's dissertation was strongly criticized by Niels Bohr and Léon Rosenfeld and was initially poorly received among physicists. More than a decade later it was revived by Bryce DeWitt as part of his interest in the use of quantum theory in studies of cosmology and he rechristened it as the "many-worlds" interpretation. From then on it received ongoing increasing attention. In the mid 1980s, after Aspect's experiments, John Bell began to publicize alternative interpretations which were realist and compatible with standard quantum mechanics, an attempt Bell called the quest for "beables" instead of "observables" which were at the core of the usual interpretations of quantum mechanics. Indeed, either the interpretations by Bohr or by von Neumann were theories where observables were the key assumptions. Bell expressed a liking towards Bohm's hidden variables but also drew attention to the existence of Everett's interpretation.[18]

Bell was also responsible for stimulating another line of approach to the quantum riddles. This was related to introducing changes in Schrödinger equation in order to deal with the collapse of quantum states during measurements. This was obtained through the introduction of non-linear and stochastic terms in the Schrödinger equation. Thus this approach is not, strictly speaking, a new interpretation of quantum mechanics; indeed it is a different theory. However, this stochastic term does not lead to different predictions for microsystems with few degrees of freedom but it is able to explain the absence of superposition of quantum states in the description of macroscopic systems. As Bell wrote, in this approach, "the [Schrödinger's] cat is not both dead and alive for more than a split second." The proposal was initially developed by the Italian physicist Giancarlo Ghirardi in collaboration with Alberto Rimini and Tullio Weber, which explains why it is sometimes called the GRW approach. Through Bell's support this team was joined by Philip Pearle, who was working along similar lines. As we have seen in the previous chapter, Bub's doctoral dissertation, in the mid-1960s, supervised by Bohm, had been a step in this direction.[19]

Everett's work and Ghirardi and colleagues' were not the only novelties on the scene of quantum mechanics research in the 1980s. Other interpretations, such as the "consistent histories" and the stochastic interpretations were suggested or gained momentum. In addition, two different subjects were arousing interest among physicists. The study of decoherence, a term which gained currency later, meant

---

[18]On Bell's works, see Bell (2004). On Everett's work and its reception, see the work by Stefano Osnaghi, Fabio Freitas, and Olival Freire in (Freire Junior, 2015, 75–139). Everett's original papers are in Everett et al. (2012).

[19]Bell (2004, 204), Ghirardi et al. (1986) and Ghirardi (2002).

to understand how idealized isolated quantum systems were indeed not isolated as they could interact at least with a few photons or atoms surrounding them. Technically this meant the Schrödinger equation could not be fully applied and that these systems needed approximations in order to be studied. The other subject was the first interest in the use of concepts from the foundations of quantum mechanics: to study storage, encryption and transmission of information.[20] All these novelties were putting David Bohm under pressure to reconsider his early thoughts on quantum mechanics. Bohm's lasting and ongoing quest to understanding the quantum, through the 1980s, however, only appeared in print in the book *The Undivided Universe— An Ontological Interpretation of Quantum Theory*, a major work co-authored with Basil Hiley, which appeared as a posthumous publication in 1993. Before commenting on this book, let us divert our conceptual thread to consider Bohm's wide and increasing recognition and the ultimate outcome of his battle to regain his American citizenship.[21]

6.4   Bohm's Rising Recognition

In the mid 1980s Bohm began to receive recognition of his work among physicists. It is not uncommon for a scientist to receive homages when their seventieth anniversary is approaching but in his case the accolades outnumbered the more usual recognitions. Let us cite and comment on the Festschrift *Quantum Implications—Essays in Honour of David Bohm*, special issues in the journal *Foundations of Physics*, nomination for the Nobel prize, and election as a Fellow of the Royal Society.

The Festschrift honoring Bohm's 70th birthday was an initiative of Basil Hiley, his long-time collaborator, and F. David Peat, a physicist and science writer with whom Bohm had begun to collaborate on works on the philosophy of science, having published in the same year the book *Science, Order and Creativity*. The volume was impressive for different reasons. Firstly for the scientific credentials of many of the authors, which included Nobel Prize laureates at the time the book appeared, Ilya Prigogine, Maurice Wilkins, and Richard Feynman, another who would win the 2003 Nobel Prize, Anthony Leggett, and a number of other distinguished scientists, such as John Bell, Roger Penrose, David Pines, Yakir Aharonov, Bernard d'Espagnat, and Jean-Pierre Vigier. Secondly for the intrinsic quality and organization of the chapters according their content. As noted by Tony Sudbery, a mathematician from the University of York, while reviewing the book, "the essays have been neatly arranged by Basil Hiley and David Peat to form a continuous fantasia on the themes that have concerned Bohm throughout his life as a physicist."[22]

---

[20]On these alternative interpretations and the early works on decoherence and quantum information, see (Freire Junior 2015, 287–338).

[21]Bohm and Hiley (1993).

[22]The Festschrift is Hiley and Peat (1987). *Science, Order and Creativity* is Bohm and Peat (1987), Sudbery (1988).

In addition to an intellectual biography, by the editors, and an autobiography, the book covers Bohm's early interests. Eugene Gross and David Pines wrote on collective phenomena such as plasma and metals; Anthony Leggett and T. D. Clark on the foundations of quantum mechanics, particularly dealing with macroscopic systems and quantum mechanics; Roger Penrose, on gravity and consciousness; and John Bell, Bernard d'Espagnat, Richard Feynman, Geoffrey Chew, and Henry Stapp, on different aspects of foundations of quantum mechanics. Bohm's early causal interpretation was also presented by Jean-Pierre Vigier and colleagues, from their point of view, distinct somewhat from Bohm's. Studies on time in science were then presented by Ilya Prigogine, Yves Elskens, Yakir Aharonov, David Albert, and David Finkelstein. The mathematical treatment to be followed in order to develop the general ideas of implicate order were explained by the mathematician C. W Kilmister, who presented the connections between the Clifford algebras and Hamilton's representation of quaternions, and by Fabio Frescura and Basil Hiley, who exploited the relationship between the mathematical entities (four-component spinors) used to describe spins in Dirac's equation and their mathematical counterparts (minimal left ideals) in the Clifford algebras. Robert Rosen, H. Fröhlich, and Maurice Wilkins dealt with physics and biology; while the first and the second were concerned with the epistemological assumptions in the two sciences the third visited the topic of the relationship between complementarity and concepts of molecular biology (Fig. 6.4).[23]

The Festschrift also reflected the wide influence of Bohm's views beyond the circle of physicists. This influence he had been building in the previous two decades through his more philosophical works, the proposal of wholeness and implicate order, which was seen by many, Bohm included, as a new paradigm with far reaching applications. This influence also resulted from his ongoing dialogues with Krishnamurti. It is noticeable that while extending the reach of his views beyond physics, Bohm was following some of the founding fathers of quantum mechanics in a similar manner, above all Niels Bohr, who thought the epistemological implications of this new physical theory should not be limited to physics itself.[24] Thus, Bohm's Festschrift included a number of different chapters dealing with as diverse subjects as neurology, psychiatry, psychology, philosophy, theory of art, aesthetics, meditation, and eastern thinking. The authors of these subjects had been dialoguing with Bohm's works, mainly with Bohm's more philosophical idea of implicate order. The authors were B. C. Goodwin, Alan Ford, Karl H. Pribram, Gordon Globus, Montague Ullman, David Shainberg, John Briggs, and Renée Weber. Sudbery, the reviewer we have already mentioned, was cautious about the last contributions, stating "the meaning and significance of this part of the book remain, to this reader at least, appropriately implicate." Sudbery's careful comment, in an otherwise laudatory review, reflects the general difficulty in accessing the fecundity of ideas in fields far from the domains where they were conceived and were thought to be valid. This is further true when the original ideas are transported as analogies or metaphors from one field to another. With Bohm's ideas of wholeness and implicate order was no different. Let us take a

---

[23] All papers are in Hiley and Peat (1987).

[24] For a review of some of these attempts, see (Joaquim et al. 2015).

Fig. 6.4 Festschrift dedicated to David Bohm on his 70th anniversary. Four Nobel laurates (Prigogine, Feynman, Wilkins, and Leggett) were among the authors. *Credits* Reproduced with kind permission of Routledge

specific case to illustrate the problem. Pribram was a distinguished researcher in the field of neuroscience and supported what he called the holonomic model of brain. In his contribution to Bohm's Festschrift, he described how he looked for a basic understanding of quantum mechanics, how he came across Bohm's works, and acknowledged Bohm for "providing the inspiration to pursue these ruminations and to give substance to them." For Pribram, the motivation to look for inspiration in physics had appeared when he was thinking about how to deal with certain events in the brain:

either represent them as wave forms or as statistical aggregates. He thought there could be some similarity with wave particle duality in quantum mechanics. However, a closer inspection of Pribram's contribution reveals he was inspired mostly by the idea of holography. Furthermore, he tells how through interactions with physicists at Berkeley (Henry Stapp and Geoffrey Chew) he got to know how widely the mathematical tool of Fourier transform is used in physics, included in Denis Gabor's works on holography, which awarded him the 1971 Physics Nobel Prize. The remaining problem is that holography is not a quantum phenomenon and Fourier transform is a mathematical resource used in both classical and quantum physics. Thus, it does not matter how useful interaction with Bohm's ideas may have been, and possibly was, one cannot strictly account this case as a case where quantum mechanics ideas have been influential on Pribram's work. The same is valid for the influence of Bohm's ideas on quantum mechanics on Pribram's investigations. For Bohm's own ideas, holography was just an inspired analogy for him to build the idea of wholeness and order, and just one of the inspirations, nothing more than this.[25]

It is noticeable that such a Festschrift might have fared better if it had been expanded to include more authors, in physics and philosophy of physics, who had been interacting with Bohm's ideas on the interpretation of quantum mechanics. Conspicuous absences include among others John Clauser, Jeffrey Bub, his former student, Abner Shimony, James Cushing, Franco Selleri, Harvey Brown, and Michael Redhead. Still, no Brazilian author, with whom Bohm worked during his exile, appeared in the book.

In the following year, 1988, *Foundations of Physics*, dedicated 6 special issues, with about 46 authors, to Bohm's seventieth anniversary. This journal was created in 1970 to deal mainly with issues concerning the foundations of quantum mechanics and Bohm had served on the editorial board. Authors included Max Jammer, Jeffrey Bub, H-D Zeh, Mark Semon, John Taylor, Peter Holland, Jean-Pierre Vigier, Stanley Gudder, Franco Selleri, Anton Zeilinger, Brian Josephson, L. P. Horwitz, R. I. Arshansky, A. C. Elitzur, John Cramer, F. David Peat, Alan Ford, Fabio Frescura, Geoffrey Chew, Henry Stapp, R. W. Tucker, Chris Dewdney, K. Kong Wan, Anthony Leggett, O. Costa de Beauregard, W. Schleich, H. Walther, John A. Wheeler, P. T. Landsberg, Leon Cohen, A. O. Barut, M. Božić, Z. Marić, Elida de Obaldia, Abner Shimony, Frederick Wittel, Shimon Malin, Lipo Wang, R. F. O'Connell, Berthold-Georg Englert, Julian Schwinger, Marlan O. Scully, Nicola Cufaro-Petroni, Philippe Gueret, Steven Rosen, and A. Kyprianidis. Jammer's paper is particularly noticeable for his incursion into the background to Bohm's 1952 papers on the hidden variables interpretation.[26]

Nominations for the Nobel Prize are only revealed 50 years later, thus it is too early to investigate Bohm's full records in the Nobel Archives. Even so, we know Bohm was nominated jointly with David Pines for the prize in 1958. The nomination

---

[25]Pribram (1987).

[26]The issues were in the volume 18, issues 7–18, 1988. On the creation of Foundations of Physics, see Freire Junior (2004, 1754). A nice picture of Bohm was included and he was invited to continue to serve on the journal's editorial board, according to letter from the editor, Alwyn van der Merwe, 10 Oct 1988, Folder A. 128, Bohm Papers. Jammer (1988).

was done by the Japanese physicist H. Nakano as an acknowledgment of the wide impact of the collective description used in plasmas and metals.[27] However, from some hints in the Bohm Papers, deposited at Birkbeck College, London, we know that in 1989 Basil Hiley, Bohm's long-standing assistant there, was asked by his head of department to send some background information in order to support Bohm's nomination for the Nobel Prize for quantum non-locality. While the Nobel Prize is awarded for single discoveries and not for the full work of a scientist, Hiley chose the latter and indicated three fields where Bohm had given major contributions: plasma, the Aharonov-Bohm effect, and the reinterpretation of quantum mechanics.[28]

The following year, on March 15, 1990, Bohm was elected Fellow of the centenary and prestigious Royal Society. Citations for the Royal Society can be wider and are public, thus we know Bohm's citation included his works with Foldy "on the theory of synchrocyclotrons;" works with Gross and Pines on plasma oscillations and electrons in metals; the introduction, through the work with Pines, of the "random phase approximation" showing "how interactions between individual electrons give rise to collective oscillations;" work with Aharonov on how the phase of the wave function is "influenced by electromagnetic fields in unpenetrated regions," which "gave rise to a number of beautiful experiments, and stimulated the development of gauge field theories." Still with Aharonov, "the elucidation of the 'fourth uncertainty principle." The citation even includes Bohm's two 1952 papers on "a hidden variables interpretation of quantum mechanics," which "have become classics," as "they disposed, by counterexample, of the famous 'proof' by von Neumann of the impossibility of such interpretation, and so cleared the way for the discussion of the really novel feature of quantum mechanics, its nonlocality, which led to the famous Aspect experiment." Finally, Bohm is cited as "the author of excellent textbooks on quantum theory and on relativity, and of several books on the meaning of modern physics which have attracted wide attention." Congratulations letters indicated that it was a long overdue homage. Indeed, such congratulatory letters came from Roger Penrose, Brian Josephson, and Abdus Salam, the latter two Nobel Prize winners. According to Salam's cable, "this is a fellowship which is well overdue."[29]

6.5   Bohm's Philosophy of Science

In the late 1980s Bohm extended his previous philosophical ideas to encompass a full philosophy of science. Indeed, as presented in the first book related to this subject, Bohm's ideas "are central to what has become known as 'the holographic

---

[27]See: https://www.nobelprize.org/nomination/redirector/?redir=archive/. Accessed on 22 Jan 2019.

[28]Basil Hiley to Sessler, 9 January 1989. Folder A172, David Bohm Papers, Birkbeck College, London.

[29]See citation at: https://collections.royalsociety.org/DServe.exe?dsqIni=Dserve.ini&dsqApp= Archive&dsqDb=Catalog&dsqSearch=RefNo==%27EC%2F1990%2F03%27&dsqCmd=Show. tcl, accessed on 14 Nov 2018. Congratulation letters, including Abdus Salam's cable on 22 Mar 1990, are in Folders A.129 and A.174, Bohm Papers.

paradigm'. These ideas […] have provided a new way of understanding a great many phenomena ranging from some of the problems of quantum physics to health care, social organization, religion, and the workings of the human mind itself." While it is beyond the scope of this book to carry out a thorough assessment of this philosophy, some of it and of its reception will be presented in order to convey a more comprehensive picture of Bohm as an intellectual. The key question in Bohm's philosophy of science was the quest for freeing creativity. This was considered the holy grail to cope with obstacles hindering the development of science and the improvement of society. The previous idea of a criticism of the Kuhnian model of 'normal science' as a fossilization of science, thus denying the continuous appearance of new conceptual schemes, was brought together with a criticism of fragmentation. Normal science and fragmentation stifle creativity. In order to liberate creativity scientists should look for totalities and different kinds of orders. This conceptual framework came with a method—dialogue—to provoke creativity. Donald Factor, a businessman interested in the arts who has been attracted to Bohm's ideas through reading *Wholeness and the Implicate Order*, was instrumental in organizing one of the first meetings where this method was implemented as well as in publishing its proceedings. According to a reviewer, "David Bohm believes that dialogues held among groups of 20–30 people may save us all through their transforming effects on the culture." According to Bohm, himself, "A form of free dialogue may well be one of the most effective ways of investigating the crisis that faces society, and indeed the whole of human nature and consciousness today. Such a form of free exchange of ideas and information is of fundamental relevance for transforming culture and freeing it of destructive misinformation, so that creativity can be liberated." Most of these ideas were systematized in collaboration with the physicist and science writer F. David Peat and were encapsulated in the book *Science, Order and Creativity* (Bohm and Peat 1987). Bohm also scattered these ideas in a number of different talks and conversations. Some of these talks predate the publication of the book, the earliest dating from the late 1960s. Many of these talks were published as books when Bohm was still alive, others were posthumously published.[30]

In the book *Science, Order and Creativity* Bohm resumed the comparisons between Western and Eastern cultures dedicating a section to how consciousness is conditioned in each tradition. This time, however, his conclusions were slightly different from those in the previous book, *Wholeness and the Implicate Order*. In *Wholeness*, dealing with the themes of fragmentation and wholeness, Bohm's thoughts had been clearly sympathetic to what he saw as the manner the Orient dealt with the notion of measure. In *Science, Order and Creativity*, Bohm and Peat revised diverse eastern traditions concerning the manner in which they deal with consciousness and took an equidistant stance, writing, "what is clearly needed in East and West is

---

[30]Reference to the holographic paradigm is by Donald Factor in the Introduction of Bohm (1985). Citations from *Utne Reader*, March/April 1991, pp. 82–83. [These are excerpts from an interview conducted by John Briggs and published in New Age Journal, Sep/Oct 1989]. Copy at Folder B.80, Bohm Papers. Bohm's books include Bohm (1985, 1994, 1996, 1998, 1999 and 2002). Bohm (1998) includes a preface by the American Indian Leroy Little Bear; Bohm (1985) has an introduction by Donald Factor; and Bohm (1994, 1996, 1998 and 2002) were edited and introduced by Lee Nichol.

a creative surge of a new order. Such a surge will not be possible while humanity goes on with its current fragmentation, represented by the extremes of Eastern and Western cultures." Furthermore, they concluded, "East and West could move forward toward a broad 'middle ground,' between Western dynamism and Eastern suspension of outward activity, as well as between the timeless and the temporal orders, the individual and the social orders, with the cosmic order on one side and the social and individual orders on the other."[31]

The reader of this biography may easily relate many of these themes with older intellectual concerns of Bohm. The defense of creativity was Bohm's counterpart to the poor reception his alternative interpretation of quantum mechanics received among his fellow physicists while creativity and dialogue as remedies for societal issues were the continuity of his long-lasting social consciousness. The sociologist Trevor Pinch identified these connections, while not always agreeing with Bohm's ideas. According to Pinch, reviewing *Science, Order and Creativity*, "Bohm and Peat are pessimists about modern life and culture. They call for a resurgence in creativity as the only way to avoid the overwhelming problems of war, poverty, environmental catastrophe and the like facing the planet. This book is an investigation of creativity and, more to the point, what blocks it." Furthermore, Pinch considered "if science in its present form is about anything, it is about bringing agreements. […] too much creativity can be a bad thing, as in the case of the interpretations of the quantum theory. The more the ideas have proliferated, the more confusion there seems to have been." This last conclusion is surprising if we recall that ten years before Pinch had studied the challenge Bohm's ideas had represented to those of von Neumann's as a case for the sociology of science. In his study one would hardly anticipate that Pinch thought this way about the diversity of interpretations of quantum mechanics.[32]

The reception of these works was analogous to that of the book *Wholeness and the Implicate Order*. While the wider public, including a number of psychologists and biologists welcomed these ideas; most scientists, mainly physicists, were indifferent, skeptical or even overtly critical of them. Let me cite—in this sense—two reviews of *Science, Order and Creativity*. P. W. Atkins, a physical chemist from University of Oxford, criticized the book for its vagueness of certain concepts, different concepts of order, its unrealistic solutions for science and society, and its "Eastern" flavor. Lee Dembart, an editorial writer for the *Los Angeles Times*, concentrated his criticisms on what he perceived as vague and unrealistic propositions and the praise of Krishnamurti's ideas, which Dembart saw as an embracing of Zen Buddhism.[33] This adverse reception contributed to create among those who were less familiar with Bohm's ideas an identification between Bohm's physical ideas and his wider philosophical speculations. Only after Bohm's death, as we will see in the next chapter,

---

[31] For the "Résumé of discussion on western and eastern forms of insight into wholeness," see Bohm (1980, 25–33). For "The Responses of East and West to the Conditioning of Consciousness," see Bohm and Peat (1987, 255–260).

[32] Trevor Pinch, Opening Dialogue, *Times Higher Education Supplement* [1989], copy at Folder B.80, Bohm Papers.

[33] Atkins (1989) and Dembart (1988).

did this identification begin to wane. Furthermore, Bohm was not the first, and certainly not the last, scientist whose scientific ideas were presented *pari passu* with broader philosophical, political, or religious ideas. To discern them should be a duty of scientists at large and not only of historians and philosophers of science.

6.6 The American Citizenship Eventually Recovered

In the twilight of the Cold War, Bohm eventually won the right to recover his American citizenship after living more than 30 years as a Brazilian citizen. The right to citizenship in the US has a long juridical history, which we are not going to resume here, being enough to recall that before the Civil War slaves were not entitled to citizenship. A turning point in the history relevant for Bohm's case happened in 1967 as a consequence of the US Supreme Court decision concerning the case *Afroyim versus Rusk*. When Beys Afroyim won the right to keep his American citizenship in the Supreme Court after having voted in a foreign country, the legal decision implied that the loss of citizenship would require a voluntary renunciation to this right. In the years following this decision, the US Judiciary and Executive government consolidated the US had no right to cancel a citizenship unless the subject had consciously and voluntarily asked for this. However, the Supreme Court decision did not lead to a sudden change, as noted by Peter Spiro, Professor of International Law at the University of Georgia: "although the decision on its face appeared to preclude unwilling expatriation in all cases, it would take more than twenty for that vision fully to take hold." Still, according to Spiro, "in the meantime, expatriation without the individual's overt consent, though more closely scrutinized, remained an option in some cases."[34] In addition to this rather slow change, there was Bohm's previous experience. Indeed, lack of hope caused by the previous failed attempts meant that Bohm took time to recover his citizenship once more. It was after the ascension of Mikhail Gorbachev to power in the Soviet Union, in 1985, and the expectations of an end to the Cold War it triggered, that Bohm decided once again to apply to recover his US citizenship and obtain an American passport.

Through a legal process, Bohm's lawyer argued Bohm did not want to renounce his American citizenship. He used Bohm's letters to Einstein written from Brazil, in which it was clear that Bohm did not intend to give up American nationality, and that he had applied for Brazilian citizenship only in order to get a passport to travel abroad as part of his professional activities. Indeed, in these letters there was no record of Bohm's willingness to abandon his US citizenship and to adopt Brazilian citizenship when he applied for Brazilian citizenship in 1954. Bohm won the case in 1986. He joyfully received the official communication about the restoration of his citizenship: "Dear Dr. Bohm. I am pleased to inform you that the Department of State has today notified the Embassy that your citizenship case has been reconsidered. It has now been determined that your naturalization which took place in Brazil, in November

---

[34]On the *Afroyim versus Rusk* case, see Spiro (2005), particularly p. 159.

1954, was an involuntary act. Consequently your loss of United States citizenship has been overturned; and the Certificate of Loss of Nationality that was initially prepared has been vacated." However, victory came too late. According to Bohm's letter to his longtime friend, Hanna Loewy, he had no income to live in the US as a retiree: "I cannot see how I could settle there permanently, because my pension could not be adequate for this." In Cold War times, keeping dignity came at a high price.[35]

6.7   Updating the Quest for Understanding the Quantum

In the mid 1980s times were ripe for Bohm to visit his 1952 causal interpretation once again and try to put it in line with his 1970s implicate order approach. In addition to the events we have commented on in the first three sections of this chapter (appearance of computer-produced pictures of quantum potential and paths, reception of the book *Wholeness and the implicate order*, Alain Aspect's experiments, and the rising interest in alternative interpretations), there was the strong support John Bell was bringing to the causal interpretation, or, as he called it, the pilot wave interpretation. Furthermore, as never before, the time was ripe for research on the foundations of quantum mechanics. The prevailing spirit was well grasped by John Maddox, an old colleague of Léon Rosenfeld—the most vocal defender of Bohr's thoughts from the 1950s to the 1970s—and influential editor at *Nature*. Maddox provocatively entitled his article with the question "Licence to slang Copenhagen?" He opened saying that "the drum-beat of subdued discontent with the standard interpretation of quantum mechanics has become almost deafening." After remarking that "not that serious interest in the meaning of quantum mechanics is all that news," he went on to conclude "the novelty in more recent articles on the meaning of quantum mechanics is that people write as if they now have a licence to be puckish about Bohr and all his works."[36]

6.7.1   Early Attempts to Combine Hidden Variables and Implicate Order

David Bohm, who had never asked permission to criticize what he called the "usual formulation" of quantum mechanics felt obliged to participate in grand style in this new intellectual ambiance. Furthermore, he needed to deal with questions such as:

---

[35]Edward Gudeon to Ehud Benamy, 11 Feb 1986, Folder A.123. Bohm recovered Einstein's letters helped by Lilly Kahler and Hanna Loewy according to documents in Bohm Papers, Folders C.28 and C.41. Richard Haegele—American Consul in London—to David Bohm, 11 Feb 1986, Folder A.123. Bohm to Hanna Loewy, 3 March 1986, Folder C.41. All documents in Bohm Papers, Birkbeck College, London.

[36]Maddox (1988). For an analysis of the changes in the physics community mood towards the research of the foundations of quantum mechanics, see Freire Junior (2015).

what is the problem, if any, with his 1952 causal interpretation? In particular, what is the problem, if any, with the quantum potential? Still, how to make the implicate order approach compatible with the causal interpretation? The most comprehensive answer to these questions would appear in his magnum opus, the book *The Undivided Universe*. However, many of the results presented in the book had been previously prepared, particularly in a couple of papers he had published with Basil Hiley and Pan N. Kaloyerou. Let us thus see the novelties in these articles. The first paper, authored by Bohm and Hiley, is an improved presentation, with some changes, of Bohm's 1952 causal interpretation, valid for non-relativistic systems while the second, which counted on his doctoral student Pan Kaloyerou as a co-author, is an extension of the previous ideas for quantum field theories, particularly for bosonic fields.[37]

The first paper provides a few novelties in comparison with previous reviews of the causal interpretation. The main novelty concerns how Bohm came to see the quantum potential. While in 1962 he considered one should not "be satisfied with accepting such a potential in a definite theory," a position he reiterated in 1980, when he republished this 1962 paper as a chapter of the book *Wholeness and the Implicate Order*, now he saw the quantum potential from an entirely new perspective. Bohm and Hiley still considered the quantum potential not to be equal to other physical potentials, as its main feature was not to bear energy, but rather, they considered it the carrier of information, what they called "active information." They noted that this potential "depends only on the form of the wave function and not on its amplitude," thus "its effect does not necessarily fall off with the distance." This was the key to accepting the quantum potential as the explanation for the now well set quantum nonlocality or entanglement. As a consequence, "a system may not be separable from distant features of its environment, and may be non-locally connected to other systems that are quite far away from it." According to them, 'the basic idea of active information is that a form having very little energy enters into and directs a much greater energy. The activity of the latter is in this way given a form similar to that of the smaller energy. It is therefore clear that the original energy-form will "inform" (i.e. put form into) the activity of the larger energy."[38]

Bohm and Hiley's argument was strongly analogical. They accepted quantum non-locality as a well-established physical effect and wanted to keep the quantum potential as it seemed to be an indispensable part of the hidden variables interpretation. However, in order to obtain an explanation for this potential they appealed to analogies with cases where a wave carried information while the source of energy for movement was elsewhere. Let us follow Bohm and Hiley to see how these analogies worked[39]:

> As an example, consider a radio wave whose form carries a signal. The energy of the sound that we hear from our radio comes, however, not from this wave but from its power plug or batteries. This is essentially an "unformed" energy, that takes up the form or information carried by the radio wave. The information in the radio wave is, in fact, potentially active

---

[37]The two papers are Bohm and Hiley (1988), Bohm et al. (1988).

[38]Bohm and Hiley (1988).

[39]Bohm and Hiley (1988, 327).

everywhere, but it is actually active, only where and when its form enters into the electrical energy within the radio. A more developed example of active information is obtained by considering the computer. The information content in a silicon chip can similarly determine a whole range of potential activities which may be actualized through having the form of this information enter the electrical energy coming from a power source. Which of these possibilities will in any given case be actualized depends on a wider context that includes the software programmes and the responses of the computer operator at any given moment.

As the quantum potential does not necessarily fall off with distance and through it a system may be non-locally connected to other systems that are quite far away from it, the quantum potential became the main illustration of Bohm's wholeness, thus connecting the 1952 causal interpretation with his later wholeness and implicate order approach.

Attractive as this explanation may be, and it is, it is not devoid of problems. Usual waves conveying information are waves of a physical type. Electromagnetic waves and sound waves, for instance. Thus the idea of the quantum potential as the carrier of active information still requires an explanation for its physical nature. Furthermore, when Bohm and Hiley wanted to discuss the attainment of the classical limit, that is, when a system is no longer described by quantum mechanics and may from this point on be described by classical physics, they suggest doing this "by neglecting the quantum potential," thus leading one to the Hamilton-Jacobi equation. As they argued, "such neglect will be justified when the quantum potential term is actually small compared with the other terms in the Hamilton-Jacobi equation." Thus it reappears a comparison between the quantum potential and the potentials which are known so far. However, leaving this difficulty aside, this approach has the merit, if the quantum potential is accepted, of explaining the classical limit in a more reasonable manner. According to them, "roughly speaking, then, classical behaviour is approached, when the classical potential dominates over the quantum potential; and when this is not so, we obtain typical quantum behaviour." This explanation is thus different from the explanation you may find in some quantum physics textbooks that the classical limit is attained when Planck's constant reaches zero. This usual explanation is an artificial one because if Planck's constant is a constant of nature, how can it go to zero? According to Bohm and Hiley, "the approach to the classical limit has therefore nothing to do with setting $\hbar = 0$, as Planck's constant is always actually the same, while the relative importance of the terms in the equations is what may change with conditions." This conundrum is an impressive example of the difficulties still present in the foundations of quantum mechanics. The appeal to the quantum potential may not be convincing but it may be positively compared to other explanations which also bring some artificiality to the understanding of quantum mechanics.[40]

The second paper, which was part of Pan Kaloyerou's doctoral work, is an extension of the causal interpretation to deal with the electromagnetic field itself, thus approaching quantum field theory. They departed from Bohm's 1952 initial insight in this domain but acknowledged that this previous work contained formal

---

[40]Bohm and Hiley (1988, 341).

complications which hid its physical features; thus they decided to work on a simplified model, a scalar massless bosonic field. They were able to reproduce a number of results from quantum electrodynamics, including interferences and the Fock states which had been part of Roy Glauber's seminal work with quantum optics. In the approach adopted by Bohm, Hiley and Kaloyerou, the field is always continuous but due to the non-linearity and non-locality present in the equivalent quantum potential, the field manifests itself in the interaction with atoms as a discrete, particle-like, physical reality. The paper also suggests, but does not develop, an extension of the causal interpretation to fermions.[41]

6.7.2 Undivided Universe—The Synthesis

Twenty-five years later the 400-page book *The Undivided Universe—An ontological interpretation of quantum theory* remains a good introduction to the foundations of quantum mechanics and the diversity of their interpretations written from a certain perspective, mostly that of hidden variables as they were developed and interpreted by David Bohm and Basil Hiley. It is both a masterwork for its clarity, and a tour de force for its broad scope. Today it may be compared, for instance, to the books written by Detlef Dürr and Stefan Teufel, in 2009, *Bohmian Mechanics—The Physics and Mathematics of Quantum Theory*, and by Peter Holland, in 1993, *The Quantum Theory of Motion: An Account of the de Broglie-Bohm Causal Interpretation of Quantum Mechanics*, which are books with similar goals but with their own and distinct perspectives. Still, Bohm and Hiley's book may be compared, in scope, to the more recent, from 2012, *Do we really understand quantum mechanics?* which was written by Franck Laloë to introduce physicists to the plethora of foundational and interpretational issues in quantum theory, while Laloë's book is so far the most accomplished attempt to introduce these issues without bias.[42]

The book *Undivided Universe* begins discussing interpretations of quantum mechanics contrasting the epistemological view, mainly represented by Niels Bohr, and the ontological view supported by Bohm and Hiley. They exploited the distinction between Bohr's complementarity principle and Heisenberg's view of potentialities as well as von Neumann's idea of quantum states to state that while Bohr's is the most strictly epistemological interpretation, von Neumann's view concedes more to an ontological view because "it assumed the quantum system existed in a certain quantum state". Then they discussed if Bohr's views are inevitable. They took the assumptions on which Bohr had based his conclusions—indivisibility of the quantum of action and unpredictability and uncontrollability of quantum phenomena—and challenged them. On a deeper level the action may be continuous, they argued, and recalled that Bohr's identification between predictability and determinism is no longer valid after the discovery of a number of systems, unstable and chaotic,

---

[41] Bohm et al. (1988).

[42] Bohm and Hiley (1993), Dürr and Teufel (2009), Holland (1993a) and Laloë (2012).

where the lack of predictability does not mean that such systems are not determinist. Bohm and Hiley added that Bohr had philosophical assumptions—positivist philosophy—which led him to leave aside the idea of an independent physical reality. After recalling von Neumann's proof against the possibility of changing quantum mechanics through the introduction of additional (hidden) variables had been challenged by Bohm and Bell, they come to the conclusion that "this book presents in essence the first complete ontological interpretation that has been proposed" while acknowledging that "there have been several other ontological interpretations since then," which will be discussed and compared with their own interpretation (Fig. 6.5).[43]

After these preliminary philosophical considerations, the authors presented their ontological interpretation, calling it the causal interpretation, a name reminiscent of the scientific and philosophical battles of the 1950s. They successively dedicated chapters to the one-body system, many-body system, transition processes, measurement, nonlocality, and the classical limit of the quantum theory. While most of the content is based on previous publications, there is a myriad of new phenomena and applications which had not been dealt with in the first presentations of the causal interpretation in the early 1950s. Treated from the point of view of this interpretation they include topics such as the Aharonov-Bohm effect, chemical bonds, superfluidity and superconductivity, delayed choice experiments, watchdog effect and Zeno's paradox, Bell's inequalities, and Bell's theorem for three particles. Many of these phenomena were unknown in the early 1950s. In several cases the mathematical treatment was illustrated by graphs first obtained using computers by Chris Philippidis and Chris Dewdney. Furthermore, there were the new concepts: the quantum potential as the bearer of active information and the expression of the quantum wholeness, along the lines of the paper they had published earlier as we have already discussed in this chapter.[44]

One of the book chapters is dedicated to the relationship between the ontological and causal interpretation and the statistics present in the usual formalism of the quantum theory. This was not a new issue as Bohm had faced it in the publication of the hidden variables interpretation in 1952, as we have seen. As they formulated the problem, in their interpretation "the primary significance of the wave function is that it is a quantum field which determines information that is active on the particles in each individual systems," however, "the wave function determines the probability density in a statistical ensemble through the relationship $P = |\Psi|^2$." Now, they had to explain "why P should tend to approach $|\Psi|^2$ in typical situations that are currently treated in physics (i.e. situations in which the quantum laws are valid)." Bohm and Hiley recovered the solution Bohm and Vigier had adopted in 1954, which had been later extended by the mathematician Edward Nelson. This was the introduction of a stochastic movements in the medium where the particles were immersed, thus leading P to approach $|\Psi|^2$ after certain short times. It should be noticed that by the time the book was finished, at the end of 1992, Sheldon Goldstein, Detlef Dürr, and Nino Zanghi had already published a more elaborate formulation for the same

---

[43]Bohm and Hiley (1993), Chap. 2, pp. 13–26.

[44]The earlier publication is Bohm and Hiley (1988).

Fig. 6.5 The Undivided Universe—An Ontological Interpretation of Quantum Theory, by Bohm and Hiley. This was the last book written by David Bohm. *Credits* Reproduced with kind permission of Routledge

problem, which we will comment on in the next chapter, however, Bohm and Hiley did not refer to their work. Finally, this chapter allowed Bohm to resume the old philosophical issue of the status of deterministic laws in physics. As they remarked, "if we abstract these chaotic motions and consider them apart from their possible causes we have what is called a stochastic process which is treated in terms of a well-defined mathematical theory." However, how to consider the status of such processes in physics? Bohm and Hiley's answer was that "we can have two attitudes to such a stochastic process. The first is that it is a result of deeper causes that do not appear at the level under discussion. The second is that there is some intrinsic randomness in the basic motions themselves." The solution they presented for this either/or question reminds us of Bohm's 1957 book *Causality and Chance in Modern Physics*, which, expressing Bohm philosophical studies of Hegel, asserted that both chance and causality may have equal foot in physical laws. According to Bohm and Hiley, "in so far as we apply the ordinary mathematical treatment, we need not commit ourselves to either attitude." We may thus conclude that while they had used the same term from the early 1950s, causal interpretation, its meaning had been extended. That is the reason for the choice of the term ontological interpretation, instead of causal interpretation, for the book's title.[45]

Bohm and Hiley next extended their treatment to deal with spin, bosonic fields as electromagnetic fields, and the issue of relativistic invariance of the causal interpretation. They revisited the first work of Bohm, Schiller, and Tiomno on the Pauli equation for spin, showed their limitations, and went on to work with this equation as a non-relativistic limit of the Dirac equation. For bosonic fields they extended the treatment presented slightly earlier in the joint work by Bohm, Hiley, and Kaloyerou, which we have already reported. On the general question of the relativistic invariance of their approach, they first showed that there is a Lorentz invariant description "of the manifest world of ordinary large scale experience" and that "all statistical predictions of the quantum theory are Lorentz invariant in our interpretation." However, it remains an issue of full compatibility between the causal interpretation and the relativistic invariance, as was pointed out by the authors. "We cannot maintain a Lorentz invariant interpretation of the quantum nonlocal connection of distant systems," and "where individual quantum processes are concerned, our ontological interpretation requires a unique frame of the kind we have described both for field theories and particle theories." Bohm and Hiley did not consider this a weakness of their approach, instead they saw it as an indication of the need to take into account "new levels of reality."[46]

The authors also compared their interpretation with other ontological alternative interpretations or approaches, such as the Everett's interpretation, decohering histories, and Ghirardi-Rimini-Weber approach. The final chapter was dedicated to the implicate order. It recollected the previous works, including those where the need for new algebras and the possibility of having space-time as an emerging feature from these algebras had been introduced. Noticeably, it is the part of the book where systematization of previous works prevailed in the exhibition of new results. As for the compatibility between the early hidden variable interpretation and the wholeness and order approach, the results were meager. Indeed, except for the suggestion of the quantum potential as the illustration of the wholeness and the implicate order, no stronger connection between the two approaches (hidden variables and wholeness) Bohm had devoted his life to was attained.

---

[45]Bohm and Hiley (1993, 193–194), Bohm and Vigier (1954), Dürr et al. (1992) and Bohm (1957).
[46]Bohm and Hiley (1993, 271).

This last chapter still includes the most daring conjecture of the entire book. After maintaining the existence of similarities between physical and mental processes, particularly through the idea of implicate order, the authors jumped to the boldest conclusion: "both are essentially the same," and "it is thus implied that in some sense a rudimentary mind-like quality is present even at the level of particle physics." Still, in a final and breath-taking leap, "each human being similarly participates in an inseparable way in society and in the planet as a whole," and "such participation goes on to a greater collective mind, and perhaps ultimately to some yet more comprehensive mind in principle capable of going indefinitely beyond even the human species as a whole."[47] No surprise such panpsychism would dismay some readers.

We may notice that there were a few blind points in Bohm and Hiley's magnum opus that only in hindsight can be clearly seen. On the one hand, there was no dialogue with the attempts to quantize gravity, an endeavor already in movement, where ideas such as discrete space-time and space-time as emergent features, as well as mathematical resources such as topology and generalized algebras were being used. On the other hand, the centrality of the concept of active information could have triggered a dialogue with the emerging field of quantum information and the attempts to rebuild the quantum theory placing the concept of information at the base of this theory, which did not happen.[48]

6.7.3 The Reception of the Undivided Universe Among Physicists

An inspection of the reviews of *The Undivided Universe* reveals that it awakened mixed feelings. It was highly praised for its breadth, clarity, and originality as a book. The reviewers also made justice to the reach of Bohm's contributions to quantum physics given that the book was posthumously published. However, the physical solutions presented in the book did not deserve similar praise. As we will see, the reviewers were cautious and unconvinced in most cases, and, in a few points, even strongly critical. Let us see some of these reviews as they are good markers of the reception of a book.

An interesting review, with which much of the scientifically skilled readership would agree, was written by the theoretical physicist John Polkinghorne. After synthesizing Bohm's ideas on the hidden variables interpretation, Polkinghorne considers the posthumous book "the fullest available account of these ideas. Much insightful discussion and detailed calculation is set before the reader." Implicit in this praise is the following conclusion: the book is noteworthy by its synthesis and pedagogical

---

[47]Bohm and Hiley (1993, 381–388).

[48]On the emergence of quantum information as a research subject, see Freire Junior (2015, 327–331). Later, Basil Hiley engaged in the analysis of the relationship between his approach and the emerging research on quantum information; see Maroney and Hiley (1999). On popular books about quantum gravity, see Smolin (2000) and Rovelli (2017).

features but does not bring any novelty, in scientific terms, able to make it distinct from Bohm's previous work. Thus the reviewer did not notice or did not value the role Bohm and Hiley attributed to the concept of active information related to the quantum potential. Polkinghorne also remarked that only the final chapter was indeed related to Bohm's wholeness approach and aptly observed that as far as this approach was concerned "the authors see the main divide between themselves and others as relating to how fundamental a role is assigned to Hilbert space, the conventional mathematics expression of the quantum formalism." According to Polkinghorne, "in contrast to their rivals, they give it no special status." In addition to compliments to Bohm's work and creativity, Polkinghorne listed his dissatisfactions. First, the old problem Bohm had faced in early 1950s: how a random state of motions could converge to the quantum probability distributions? A simple but restrictive answer is to say "the initial probability distribution is of a particular compatible kind," which is supplemented by Bohm and Hiley with the consideration that in the general cases "environmental effects would soon bring it about that it became so." Polkinghorne considered this solution "somewhat vague and not wholly satisfactorily spelled out." Second, still an old problem, the extension of this interpretation from the non-relativistic case to formulations able to reproduce results from quantum field theories. According to Polkinghorne, Bohm and Hiley's work suggests that on one hand, in the case of bosons, "there are fields but no particles", thus no "photon-like" events. On the other hand, with fermions, "there are particles but no fields." Polkinghorne considered all these solutions to seem "rather odd." Another reviewer, the philosopher John Leslie, praised Bohm as "one of this century's finest and most attractive minds," and Bohm and Hiley for their endurance in the research on the hidden variables interpretation, recalling that "over the years, Bohm and Hiley developed this theory in detail, guarding it against the criticisms which had led de Broglie to abandon it." Despite these praises, Leslie, who was a lecturer of philosophy including philosophy of science and philosophy of mind, could not accept Bohm and Hiley's boldest conjecture. For him as the book went as far as "to suggest that enfoldments have given rise to a 'collective mind' which might extend indefinitely beyond the human species," the resulting effect may have been the alienation of physicists from the subject. Thus, "no wonder so many physicists are reluctant to touch this kind of thing."[49]

Reviews from those physicists who were working on the foundations of quantum mechanics were particularly interesting. Sheldon Goldstein, a physicist from Rutgers University, baptised Bohm and Hiley's proposal "Bohmian mechanics" and presented it along the lines of derivation which were simpler and cleaner than those by Bohm and Hiley. This derivation, which had been developed by him and colleagues, was based on the same ontology; electrons are moving particles, thus with well defined position and speed, guided by waves obeying Schrödinger equation. Goldstein did not need the introduction of the quantum potential and presented considerations from statistic mechanics to the convergence of the general distribution of probabilities towards the distribution $\mid \Psi \mid^2$ given by quantum mechanics. Furthermore, as he did not need the quantum potential in his derivation, he considered Bohm and Hiley not radical

---

[49]Polkinghorne (1994) and Leslie (1994).

enough: "They formulate Bohmian mechanics in terms of the 'quantum potential,' which permits the guiding equation to be recast into a classical, Newtonian form, but at the price of obscuring the basic structure and the defining equations of the theory and of injecting an appearance of artificiality into its formulation." Goldstein acknowledged, however, that "the quantum potential plays a genuine role only in their analysis of the classical limit; elsewhere, it is the wave function and the (original) guiding equation that are relevant." Goldstein also considered the treatment of spin unsatisfactory and regretted that "recent important developments in the analysis of the origin of randomness in a Bohmian universe are unfortunately omitted." Finally, in the possibly strongest criticism, he considered that "the speculations in the last chapter about the "implicate order" failed to enhance our understanding of Bohmian mechanics; on the contrary, before the reader has had time to digest this theory, he is given the impression that it depends upon these speculations for its adequacy." These were not innocent considerations, as we will analyze in the next chapter.[50]

Another reviewer, also an expert in the domain of foundations of quantum mechanics, was milder than Goldstein in his criticism. Peter Holland, from the Université Pierre et Marie Curie, Paris, published in the same year a textbook presenting the hidden variables approach along the lines shared by him and Jean-Pierre Vigier. Holland considered Bohm's "most enduring bestowal may well prove to be his work in an area many have felt to be on the fringes of acceptable physics: the interpretation of quantum mechanics that he first proposed in 1952," and valued Bohm's theory because it has "the major advantage of conceptual clarity, and the paradoxes that afflict other views simply do not arise." However, Holland criticized the idea of the quantum potential as the carrier of active information because "it is a view that some may find unpalatable because it seems to read more into the mathematics than is actually there," and also regretted that "other points of view consistent with the basic model" were not represented. Thus, "the book does not provide a comprehensive review of the field as it now exists." Finally, Henry Stapp, from the University of California at Berkeley, who had also dedicated his attention to the field of foundations was rather favorable to the book. He considered the hidden variables and the orthodox interpretations were on a par with each other and adopted a pragmatic stance about the dispute between them: "as regards utility, the verdict is mixed, the orthodox theory is far simpler for computations, but Bohm's theory is logically cleaner and closer to simple intuitions." Stapp went on to suggest possible experiments to distinguish the two interpretations in an unforeseen field, brain and human vision. The argument was sound and went back to the early debates on Bohm's hidden variables. Werner Heisenberg, and previously Wolfgang Pauli, had criticized this approach because it did not reproduce the equivalence between position and momentum representations, which is a feature of the standard mathematical formalism of quantum theory. In Bohm's theory, the position coordinates are privileged. Stapp recalled that as the response functions of rods and cones in the retinas of our eyes depend on both the

---

[50]Goldstein (1994).

position and momentum of the detected photons, some kind of empirical distinction between the two approaches is to be expected. Unfortunately, the suggestion was vague and, as far as I could check, never become a real empirical test.[51]

6.8   Health Issues and Ultimate Result

In early 1991 David Bohm's health problems became more severe. States of depression, a long haunting qualm for Bohm, returned in such a strong manner as never before. To overcome the crisis Bohm was treated with electric shocks, not an easy decision for the family to take. From the end of 1991 and during 1992, his depression lifted and Bohm came back with full energy to work with Basil Hiley on the book *The Undivided Universe*. In January 1992, the author of this biography visited David Bohm at Birkbeck College. He was 74 years old and seemed in good health. However, Bohm was taking a number of medicines for depression and heart issues, a difficult trade-off (Fig. 6.6).

On October 27, 1992, in the early evening, while in a cab from Birkbeck College to his home in Edgware, Bohm suffered a heart attack. The cab driver called Sarah, Bohm's wife, and they ran into the Edgware General Hospital, but the staff could not

Fig. 6.6 Sarah and David Bohm, and Karl Pribram, Prague, June 1992. One of the last Bohm's pictures. *Credits* Reproduced with kind permission of Katherine Neville. The picture is at http://www.katherineneville.com/karl-pribram-and-friends/david-bohm-june-1992/

---

[51] Holland's book is Holland (1993a) and the review is Holland (1993b). Stapp (1994).

resuscitate him. Sarah's recalled that Bohm had called her a quarter past six, still at Birkbeck, to let her know he was taking a cab. Sarah's annotations reveal, "he was bubbling with excitement and full of energy. He said that things have been going so very well. 'It is tantalizing, I feel that I'm on the edge of something! ... we're getting somewhere'." Less than an hour later the half century quest for understanding the quantum was over.[52]

References

Atkins, P.W.: Wholesome ideals. Nature **338**, 28 (1989)

Baracca, A., Bergia, S., Del Santo, F.: The origins of the research on the foundations of quantum mechanics (and other critical activities) in Italy during the 1970s. Stud. Hist. Philos. Mod. Phys. **57**, 66–79 (2017)

Belinfante, F.J.: A Survey of Hidden-Variables Theories. Pergamon, New York (1973)

Bell, J.S.: On the impossible pilot wave. Found. Phys. **12**, 989–999 (1982). Reprinted in Bell, J.S.: Speakable and Unspeakable in Quantum Mechanics: Collected Papers on Quantum Philosophy. With an Introduction by Alain Aspect, pp. 159–168. Cambridge University Press, Cambridge (2004)

Bell, J.S.: Speakable and Unspeakable in Quantum Mechanics: Collected Papers on Quantum Philosophy. With an Introduction by Alain Aspect. Cambridge University Press, Cambridge (2004)

Bohm, D.: Causality and chance in modern physics. Foreword by Louis De Broglie. Routledge and Paul, London (1957)

Bohm, D.: Hidden variables in the quantum theory. In: Bates, D.R. (ed.) Quantum Theory—III—Radiation and High Energy Physics, pp. 345–387. Academic, New York (1962)

Bohm, D.: Quantum theory as an indication of a new order in physics. Part A. The development of new order as shown through the history of physics. Found. Phys. **1**(4), 359–381 (1971)

Bohm, D.: Quantum theory as an indication of a new order in physics. Part B. Implicate and explicate order in physical law. Found. Phys. **3**(2), 139–168 (1973)

Bohm, D.: Wholeness and the Implicate Order. Routledge & Kegan Paul, London (1980)

Bohm, D.: Unfolding Meaning—A Weekend of Dialogue with David Bohm. Routledge, London (1985)

Bohm, D.: Thought as a System. [With a Foreword by Lee Nichol]. Routledge, London (1994)

Bohm, D.: On Dialogue. [Edited by Lee Nichol]. Routledge, London (1996)

Bohm, D.: On Creativity. [Edited by Lee Nichol]. Routledge, London (1998)

Bohm, D.: Limits of Thought: Discussions with Jiddu Krishnamurti. Routledge, London (1999)

Bohm, D.: The Essential David Bohm. [Editor Lee Nichol, Preface by the Dalai Lama]. Routledge, London (2002)

Bohm, D.J., Hiley, B.J.: Intuitive understanding of nonlocality as implied by quantum-theory. Found. Phys. **5**(1), 93–109 (1975)

Bohm, D., Hiley, B.J.: An ontological basis for the quantum theory I. Non-relativistic Part. Syst. Phys. Rep., **144**(6), 323–348 (1988)

---

[52]Pa[ve] Buckley to Bohm, 17 Sep 1991, Folder C.7, Bohm Papers; Sarah Bohm's reminiscences, without date, Bohm Papers. I am thankful to Sue Godsell for sending me these reminiscences on 5 June 2016. Bohm's severe depression along 1991 is described in a detailed manner in Peat (1997, 303–322). Peat suggests the trigger of this last burst of depression was the publication after Krishnamurti's death of a book with problematic revelations about Krishnamurti's personal life. See Peat (1997, 305). The book is Sloss (1991).

Bohm, D., Hiley, B.J.: The Undivided Universe: An Ontological Interpretation of Quantum Theory. Routledge, London (1993)

Bohm, D., Hiley, B.J., Kaloyerou, P.N.: An ontological basis for the quantum theory II. A Causal Interpretation Quantum Fields, Phys. Rep. **144**(6), 349–375 (1988)

Bohm, D., Peat, F.D.: Science, Order, and Creativity. Routledge, London (1987)

Bohm, D., Vigier, J.P.: Model of the causal interpretation of quantum theory in terms of a fluid with irregular fluctuations. Phys. Rev. **96**(1), 208–216 (1954)

Curd, M.: Wholeness and the implicate order. Phys. Today **34**(8), 58 (1981)

Dürr, D., Goldstein, S., Zanghi, N.: Quantum chaos, classical randomness, and Bohmian mechanics. J. Stat. Phys. **68**(1–2), 259–270 (1992)

Dembart, L.: The Zen of Quantum Mechanics. In: Los Angeles Times, The Book Review, p. 2 (1988)

Dürr, D., Teufel, S.: Bohmian Mechanics—The Physics and Mathematics of Quantum Theory. Springer, Berlin (2009)

Everett, H., Barrett, J.A., Byrne, P. (eds.).: The Everett Interpretation of Quantum Mechanics: Collected Works 1955–1980 with Commentary, pp. 57–60. Princeton University Press, Princeton (2012)

Freire Junior, O.: The historical roots of "foundations of quantum mechanics" as a field of research (1950–1970). Found. Phys. **34**(11), 1741–1760 (2004)

Freire Junior, O.: The Quantum Dissidents—Rebuilding the Foundations of Quantum Mechanics 1950–1990. Springer, Berlin (2015)

Galison, P.: Image and Logic: A Material Culture of Microphysics. University of Chicago Press, Chicago (1997)

Ghirardi, G.C.: John Stewart Bell and the dynamical reduction program. In: Bertlmann, R.A., Zeilinger, A. (eds.) Quantum [Un]speakables—From Bell to Quantum Information, pp. 287–305. Springer, Berlin (2002)

Ghirardi, G.C., et al.: Unified dynamics for microscopic and macroscopic systems. Phys. Rev. D **34**(2), 470–491 (1986)

Goldstein, S.: The Undivided Universe: An ontological interpretation of Quantum Theory. Phys. Today **47**(9), 90 (1994)

Hiley, B.J., Peat, F.D. (eds.).: Quantum Implications: Essays in Honour of David Bohm. Routledge & Kegan Paul, New York (1987)

Holland, P.R.: The Quantum Theory of Motion: An Account of the de Broglie-Bohm Causal Interpretation of Quantum Mechanics. Cambridge University Press, Cambridge, England (1993a)

Holland, P.R.: Fringe physics. Nature **366**, 420 (1993b)

Jammer, M.: David Bohm and his work on the occasion of his 70th-birthday. Found. Phys. **18**(7), 691–699 (1988)

Joaquim, L., Freire Junior, O., El-Hani, C.: Quantum Explorers: Bohr, Jordan, and Delbrück Venturing into Biology. Phys. Perspect. **17**, 236–250 (2015)

Kaiser, D.: Drawing Theories Apart—The Dispersion of Feynman's Diagrams in Postwar Physics. Chicago University Press, Chicago (2005)

Kaiser, D.: How the Hippies Saved Physics: Science, Counterculture, and the Quantum Revival. W. W. Norton, New York (2012)

Krishnamurti, J.: Freedom from the Known. Harper & Row, New York (1969)

Laloë, F.: Do We Really Understand Quantum Mechanics?. Cambridge University Press, New York (2012)

Leslie, J.: The Absolute Now. London Rev. Books, **16**(9), 15–16 (1994)

Maddox, J.: Licence to slang Copenhagen? Nature **332**, 581 (1988)

Maroney, O., Hiley, B.J.: Quantum State Teleportation Understood through the Bohm Interpretation. Found. Phys. **29**(9), 1403–1415 (1999)

Park, D.: Review of wholeness and the implicate order. Am. J. Phys. **49**(8), 796 (1981)

Peat, F.D.: Infinite Potential: The Life and Times of David Bohm. Addison Wesley, Reading, Ma (1997)

Philippidis, C., Dewdney, C., Hiley, B.J.: Quantum interference and the quantum potential. Nuovo Cimento della Societa Italiana di Fisica B—Gen. Phys. Relat. Astron. Math. Phys. Methods **52**(1), 15–28 (1979)

Polkinghorne, J.: Worlds of Difference, Times Higher Education Supplement, January 28, p. 21 (1994)

Pribram, K. H.: The implicate brain. In Hiley, B.J., Peat, F.D. (eds.) Quantum Implications: Essays in Honour of David Bohm, pp. 365–371. Routledge & Kegan Paul, New York (1987)

Rosen, S.M.: David Bohm's wholeness and the implicate order: an interpretive essay. Man Environ. Syst. **12**(1), 9–18 (1982)

Rovelli, C.: Reality is Not What it Seems—The Journey to Quantum Gravity. Riverhead, New York (2017)

Said, E.W.: Orientalism. Pantheon Books, New York (1978)

Shimony, A:. Meeting of physics and metaphysics. *Nature* **291**, 435 (1981)

Shimony, A.: Search for a Naturalistic World View, vol. 2. Cambridge University Press, Cambridge (1993)

Sloss, R.R.: Lives in the Shadow with J. Krishnamurti. Bloomsbury, London (1991)

Smolin, L.: Three Roads to Quantum Gravity. Weidenfeld & Nicolson, London (2000)

Spiro, P.J.: *Afroyim*: vaunting citizenship, presaging transnationality. In Martin, D.A., Schuck, P.H. (eds.) Immigration Stories, pp. 147–168. Foundation Press, New York (2005)

Stapp, H.P.: The undivided universe: an ontological interpretation of quantum theory. Am. J. Phys. **62**, 958–960 (1994)

Sudbery, T.: Hidden agenda. Nature **331**, 26 (1988)

Werskey, G.: The Marxist critique of capitalist science: a history in three movements? Sci. Cult. **16**(4), 397–461 (2007)

Chapter 7
Epilogue

In the years following Bohm's death there was rising recognition of his contributions to physics and philosophy. While some of this recognition came through accolades and historical and philosophical works on Bohm's life and work, the most influential and lasting came through the people who had been working on Bohm's ideas. However, those working on these needed to deal with some ambiguities intrinsic to the Bohm's ideas; thus, depending on the choices each made from the menu, they went to work in different intellectual directions. Still, some who introduced new ideas to Bohm's original arsenal and claimed to be working along Bohm's lines, for instance those who coined the term "Bohmian mechanics," did not find wide acceptance among others who were working along different lines. Therefore, combining the rising recognition with the diverse manners of interpreting Bohm's intellectual legacy, a dispute emerged over the memory of Bohm's works. Let us now examine each piece of this puzzle before trying to bring them together.

The rise in Bohm's prestige is illustrated by the laudatory obituaries published after his death. The half-page obituary in *Nature*, the traditional British scientific journal, was written by the philosopher of physics, Redhead, who regretted Bohm's passing away as a "sad loss to the community of physicists, and particularly to those concerned with fundamental issues in quantum mechanics," and stated "in the longer term [while compared to the Aharonov-Bohm effect] his exhaustive investigations on unorthodox interpretations of quantum mechanics may be seen as an even more significant contribution." Horgan, a well-known science journalist, published a 4-page interview, possibly Bohm's last, in the *New Scientist*. Horgan titled it "Last words of a quantum heretic," and introduced Bohm's research on the foundations of quantum mechanics saying "no physicist was more sensitive to the [quantum] conundrums or worked harder to resolve them than Bohm," and concluded the article stating "no matter how history treats Bohm's specific ideas, this, surely, will be one of his greatest legacies: his ability to make the rest of us perceive and think differently." *The New York Times* presented Bohm as "physicist and writer on quantum theory," citing physicist N. David Mermin, from Cornell University, who said about Bohm's 1951 Quantum Theory textbook, "there was a time when that was probably

© Springer Nature Switzerland AG 2019
O. Freire Junior, *David Bohm*, Springer Biographies,
https://doi.org/10.1007/978-3-030-22715-9_7

the best of the available textbooks about quantum mechanics, which lies at the heart of contemporary physics." In addition, the newspaper's journalist Pace presented Bohm's troubles with McCarthyism recollecting how Bohm had been assured the right to remain silent: "Judge Alexander Holtzoff of Federal District Court ruled that Professor Bohm was within his constitutional rights in pleading that his answers might expose him to prosecution." Goldstein, from Rutgers University, wrote the obituary for *Physics Today*, the influential house organ of the American physics community. After summarizing Bohm's diverse contributions to plasma, metal, foundations and interpretation of quantum mechanics, Goldstein concluded his article bringing together Bohm's fight against McCarthyism and the orthodox view of quantum mechanics. The common ground, as found by Goldstein, was presenting Bohm as "a person of extraordinary commitment to principle, both moral and scientific." He then illustrated this by citing, "his refusal in 1951 to testify against his colleagues before the House Un-American Activities Committee," and the consequences which came from this attitude, and that "most of his last 40 years he was engaged in an often lonely pursuit of scientific truth, showing little regard for prevailing fashion or orthodoxy." Goldstein also took the opportunity to correct the misconception, shared by many, of the relationship between Bohm's hidden variables interpretation and the scientific meaning of Bell's theorem and its experiments. He initially provoked his fellow physicists, "it is ironic that many physicists continue mistakenly to claim that Bell's inequality has eliminated the possibility of any deterministic alternative to the Copenhagen interpretation, and in particular Bohm's." In the following lines, Goldstein clarified, "Bell was, in fact, led to his inequality by trying to understand how Bohm had succeeded in this 'impossible' endeavor."[1]

This increasing prestige was noted early on by his long-standing friend, the American physicist Melba Phillips, "it is too bad, very sad indeed, that he did not live to see how his reputation has shot up recently. His interpretation of quantum mechanics is becoming respected not only by philosophers of science but also by 'straight' physicists." Phillips wrote this on November 17, 1994, to Peat, while he was working on Bohm's biography, which would appear in 1997. In the same year, 1997, the traditional memoir for the Royal Society magazine was published by Bohm's longtime collaborator, Hiley. The acknowledgment, among his fellow physicists, of Bohm's contribution to a better understanding of quantum mechanics may be gauged through some events on three different occasions, spaced about fifteen to twenty years from each other. In the late 1950s, when Bohm's work on the causal interpretation was poorly received, thus casting doubt on its promise, few were those who, like the Scottish engineer Lancelot L. Whyte, considered the causal interpretation anything but fleeting. Rebuking Rosenfeld, Whyte considered Bohm's work comparable to Kepler's in mechanics. Later, in the early 1980s, John Bell was emphatic in attributing to Bohm's 1952 paper his motivation to the endeavor which led to Bell's theorem. "In 1952 I saw the impossible done," and "Bohm's 1952 papers on quantum mechanics were for me a revelation," were Bell's words. Still, in 1995, a sign of this late but rising prestige, appeared in the volume in honor of the centenary edition of Physical

---

[1]Redhead (1992), Horgan (1993), Pace (1992), Goldstein (1994).

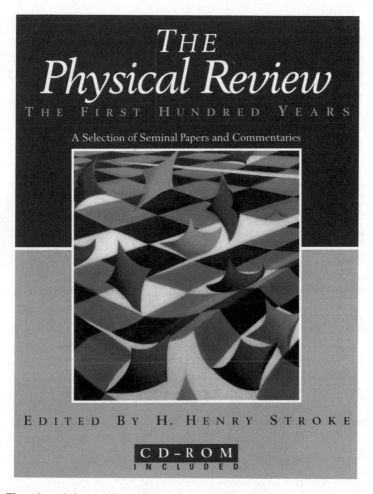

Fig. 7.1 The volume in honor of the 100th anniversary of Physical Review gathered a selection of its seminal papers

Review, the most influential American physics journal. It included commentaries and reprints of the most important papers ever published in this periodical. In the section on "Quantum Mechanics", edited by Goldstein and Lebowitz, all the papers, including Bohm's 1952 paper on the causal interpretation, concerned the foundations of quantum mechanics and a photo of Bohm opened the section (Figs. 7.1, 7.2).[2]

Bohm's prestige was also enhanced by a number of cultural, philosophical and historical studies concerning his life and work which began to appear from the mid-1990s on. Motivations for these works were diversified. Bohm's world view on

---

[2]Melba Phillips to David Peat, 17 Oct 1994, A. 22, Bohm Papers. Hiley (1997). Lancelot Whyte to Léon Rosenfeld, 8 Apr 1958, Rosenfeld Papers. Bell (1982, 1987). *Physical review* centenary book is Stroke (1995).

Chapter 14
QUANTUM MECHANICS
SHELDON GOLDSTEIN
JOEL L. LEBOWITZ

Introduction

Papers Reprinted in Book

David Bohm (1917-1992). (Courtesy of Still Pictures, London.)

Quantum mechanics is undoubtedly the most successful theory yet devised by the human mind. Not one of the multitude of its calculated predictions has ever been found wanting, even in the last measured decimal place—nor is there any reason to believe that this will change in the foreseeable future. All the same, it is a bizarre theory. Let us quote Feynman,[1] one of the deepest scientist–thinkers of our century and one not known for his intellectual (or any other) modesty, on the subject: "There was a time when the newspapers said that only twelve men understood the theory of relativity. I do not believe there ever was such a time. There might have been a time when only one man did, because he was the only guy who caught on, before he wrote his

Sheldon Goldstein is a Professor of Mathematics at Rutgers University in New Brunswick, New Jersey. He worked for many years on probability theory and the rigorous foundations of statistical mechanics. In particular, he has investigated the ergodic properties of large systems, the existence of steady-state nonequilibrium ensembles, derivations of Brownian motion for interacting particles and of diffusion and subdiffusion limits for random motions in random environments. In recent years he has been concerned with the foundations of quantum mechanics.

Joel L. Lebowitz—see biographical note in Chapter 6.

1205

Fig. 7.2 In the volume in honor of the 100th anniversary of Physical Review a full section was dedicated to foundations of quantum mechanics. Bohm's works were prominently featured

wholeness and implicate order continued to attract the attention of many. Bohm's persona in the McCarthy era came to the foreground as historians and American society at large began to revisit those dramatic times. And still, philosophers and historians focused their attention on Bohm's scientific work, his evolving epistemology, and his long quest to understand the quantum. Indeed, this type of interest had begun with Bohm still alive as sociologists of science paid attention to the bid he had done with his 1952 reinterpretation of quantum mechanics and its connections with Cold War times. The list of works is extensive and here we just refer to a sample of these numerous and diversified works: Pinch, as early as 1977, produced an analysis, based on Pierre Bourdieu's sociological framework, of the challenge Bohm presented to von Neumann's proof against hidden variables; Cross framed Bohm's causal interpretation in terms of the ideological battles launched by Soviet Marxists in the early 1950s; David Peat wrote Bohm's biography; Olival Freire Junior has given historical accounts of Bohm's causal interpretation, its early reception, his stay in Brazil, and the evolution of Bohm's ideas on quantum mechanics; Kojevnikov carried out a historical analysis of the connections between Bohm's work on plasma and his ideological commitments; Olwell studied the way Bohm was treated in the US during McCarthyism and his consequent exile; Cushing produced a philosophical analysis of Bohm's interpretation of quantum mechanics and suggested its poor reception was due to historical contingencies; Mullet studied the case of the young physicists from Berkeley, Bohm included, who became victims of McCarthyism; Pylkkänen studied Bohm's philosophy and his interaction with the artist Charles Biederman; Pylkkänen has also used Bohm's philosophical views to study the mind—matter question; Forstner studied Bohm's conversion to the causal interpretation analyzing this process in terms of Ludwig Fleck's views; and Talbot analyzed and transcribed Bohm's correspondence with Hanna Loewy, Miriam Yevick, and Melba Philipps during his exile in the 1950s. This list is not comprehensive but it gives us a clear idea of the number of different scholarly works on Bohm's life and ideas which have appeared in recent decades; this is testimony to the interest Bohm awakened among distinct researchers.[3]

7.1 The Ongoing Research on Interpretations of Quantum Mechanics

The strongest sign of the recognition of Bohm's works, however, is neither the number of homages he received nor the scholarly studies on his works. One of the most meaningful signs of the influence of a scientific work is the way other scientists are continuously working on the original work. In Bohm's case this is very clear in fields such as plasma, collective coordinates in many-body systems, and quantum effects

---

[3]Pinch (1977), Cross (1991), Peat (1997), Freire Junior (1999, 2005, 2011, 2015), Kojevnikov (2002), Olwell (1999), Cushing (1994), Mullet (2008), Bohm et al. (1999), Pylkkänen (2007), Forstner (2007, 2008), Talbot (2017).

such as the Aharonov-Bohm effect. These are fields subject to ongoing research until today. When the foundations and interpretations of quantum mechanics are concerned, the same may be said, they are subjects of continued research but with some particularities. Scientists have appropriated Bohm's early ideas in distinct and conflicting approaches giving rise to a dispute about his intellectual legacy.

These disputes are not strange to Bohm's own ideas, indeed, we may see the seeds of these disputes in his work. Let us compare, for instance, his two major last books, *Wholeness and Implicate Order* and *The Undivided Universe*. In the former the key concepts concern the ideas of order and wholeness, and the hidden variables interpretation and its quantum potential are presented as essays to show that alternative interpretations were possible but not as a scientific program to be pursued now. In addition, Bohm's ideas were presented in dialogue with Krishnamurti's philosophical views. In the latter book, the implicate order appears in a secondary role and the main bulk of the book is the defense of the hidden variable interpretation, now with the addition of interpreting the quantum potential as the bearer of active information. Furthermore, the wholeness is illustrated by the nonlocality and the quantum potential. These tensions gave rise to distinct lines of followers of Bohm's ideas. These tensions came to light as soon as Bohm passed away. Goldstein championed the camp of those who refused to accept an important role for the implicate order, considered unnecessary speculation, and defended the presentation of the hidden variable interpretations without the use of the quantum potential, as we have seen in the previous chapter. In 1993, in the long article published just after Bohm's death, the science journalist Horgan referred to Goldstein as the author of the following statement: "I've never met anyone who really understands the implicate order." As criticisms towards the implicate order grew among those who developed the so-called *Bohmian Mechanics*, to which we will refer ahead, the other line of Bohm's followers fought back. In 2013, introducing the Festschrift in homage to Hiley, the philosopher Atmanspacher noted that Hiley "does not become weary of emphasizing that many adherents of a "Bohmian mechanics", who ignore radically new features such as the implicate order and active information, are missing an important part of the ontological substance of what Bohm and Hiley stress." Atmanspacher ended with this pun. "if 'Bohmian mechanics' was the true spirit of the whole approach, both Bohm and Hiley would be truly poor 'Bohmians'!"[4]

Other researchers, such as the philosopher Redhead, also tried to keep all Bohm's ideas together and save Bohm's implicate order from the accusation of being strange to the science corpus. Thus Redhead presented Bohm's idea of order—"these [philosophical] views distinguished an explicate order, as he called it, of apparently independent entities, from an underlying implicate order involving a view of ultimate reality in which everything was bound up holistically with everything else." And added, "his critics regarded this, quite unfairly, as akin to oriental mysticism." This defense is far from being unproblematic. In fact the history of science is crowded with cases where philosophical, religious, and ideological motivations coexisted with genuine scientific developments. If mysticism is a debatable term, at least as far as

---

[4]Bohm (1980), Bohm and Hiley (1993), Horgan (1993), Atmanspacher (2013).

Jiddu Krishnamurti's thoughts are concerned, as we had seen in the previous chapter, the "oriental" intellectual influence from Krishnamurti was never denied by Bohm, and was included in the book *Wholeness and the Implicate Order*.[5] This influence did not hamper the use of the ideas of this book in scientific practice, as the work being done by Hiley and others testifies.

These disputes are, however, normal and were enhanced by the importance of Bohm's ideas for physics. They are the equivalent in science of the battle around the places of memory we see everywhere in societies. Let us therefore go on to examine the influence of David Bohm's ideas in physics through the work of their different supporters.

We group the work of Bohm's followers into three different strands. The first is directly based on de Broglie's early work, the pilot wave from the 1920s, and its improvements from Bohm's original 1952 proposal. These proposals attempt to extend the first physical models but also maintain Bohm's early philosophical commitments to determinism and realism. This line had been followed in the 1960s and 1970s, when Bohm had left it, by Jean-Pierre Vigier at the Institut Henri Poincaré and Université Pierre et Marie Curie, in Paris. One of these extensions was initiated as early as 1950s, by Vigier and Bohm, when they tried to justify the usual probabilistic interpretation related to the quantum states through the introduction of a subquantum medium endowed with random fluctuations. This led to the quest of a possible stochastic basis for quantum mechanics. Another extension was based on the search for soliton-like solutions for de Broglie's suggestion of a double-solution theory. These attempts were later systematized by Holland, who worked with Bohm and Vigier, in the textbook *The Quantum Theory of Motion: An Account of the de Broglie-Bohm Causal Interpretation of Quantum Mechanics*.

In a certain domain of physics alternative interpretations of physics have played a prominent role, the field of quantum cosmology. Cosmology became a flourishing field of research based on Einstein's theory of general relativity as this theory modified Newtonian theory of gravitation. General relativity's inception happened in the mid-1910s. However, decades later, when quantum mechanics became relevant for studies in cosmology, for instance in the behavior of the early universe, the usual interpretation of quantum mechanics became acutely problematic as the centrality of measurement or the role of an external observer making measurement on a system are clearly inadequate for a study in which the systems are the whole universe. Thus, alternative interpretations such as Bohm's approach, Everett's many worlds, and consistent histories interpretation gained traction in this field. The use of Bohm's original approach in cosmology has been championed by the Brazilian physicists, Pinto-Neto and J. Acácio de Barros. They and a few collaborators applied Bohm's causal interpretation to a number of different cosmological problems. This track has also been exploited by Valentini, who worked with deterministic hidden variables theories in the direction of relaxing the equality between distribution of hidden variables and probability distribution from usual quantum theory. For Valentini, quantum physics may be a mere case of an effective theory of an equilibrium state. Thus, he

---

[5] Redhead (1992), Bohm (1980).

has been looking for discrepancies between hidden variables and quantum theory predictions in situations of nonequilibrium. He expects to find these discrepancies in the domain of astrophysical and cosmological tests.[6]

Around the time of the publication of *The Undivided Universe* the second strand of developments based on Bohm's works gained momentum. It has been called "Bohmian Mechanics" and was inaugurated by a couple of papers co-authored by Detlef Dürr, Sheldon Goldstein, and Nino Zanghi. The papers were entitled "Quantum chaos, classical randomness, and Bohmian mechanics," and "Quantum Equilibrium and the Origin of Absolute Uncertainty." Both were published after the publication of *The Undivided Universe* and apparently Bohm did not read the preprints, or at least did not keep copies of them. The choice for this term, Bohmian mechanics, was not an innocent one. A few years later, Goldstein strived to connect their work with Bohm's original work directly, stating, in a paper co-authored with Joel Lebowitz: "we shall call Bohm's deterministic completion of nonrelativistic quantum theory Bohmian mechanics." Anyway, their titles did not reveal the entire novelty presented in these papers. Indeed, they construed Bohm's proposal in a cleaner and more elegant way. In his original paper Bohm had elaborated on analogies between Schrödinger equation and classical Hamilton-Jacobi equations, which led to an emphasis on the role of the non-classical potential that Bohm christened the "quantum potential." Bohm exploited the analogies so far that he wrote an analogous Newton's law of motion for a particle obeying quantum mechanics laws using the quantum potential in addition to the already known potentials (Fig. 7.3).[7]

Dürr and colleagues, however, departed from the two following premises: the state which describes quantum systems evolves according to Schrödinger equation and quantum particles are just particles, which means they have locations in the configuration space and they move, that is, they have a speed. The first assumption is similar to Bohm's, the difference being in how to clarify the second. Bohm had worked out analogies with Hamilton-Jacobi equations to suggest a mathematical expression for the speed, namely that the momenta of these particles were $\mathbf{p} = \nabla S(\mathbf{x})$, where S comes from writing the solution of Schrödinger equation as $\Psi = R \exp\left(\frac{iS}{\hbar}\right)$. This had led Bohm to depend on the existence of the quantum potential for his reasoning. Dürr and colleagues just assumed $\mathbf{p} = \nabla S(\mathbf{x})$ as the simplest way to attribute speed to a particle. Furthermore, Dürr and colleagues needed to justify the convergence of an arbitrary distribution of probabilities towards the Ψ obeying Schrödinger equation, a challenge also faced by Bohm. Dürr et al. took a different tack using general notions from statistical mechanics, a field where Goldstein in particular was well trained. As he recently recalled, "I never would have seen how this could all work out and come to this understanding based on typicality were it not for years, maybe even decades, of work in statistical mechanics and probability theory—work in which

---

[6]For a short description of Vigier's life and work, see Holland (1995), and for a detailed analysis of his works, see (Besson 2018). The textbook systematizing these ideas is Holland (1993). A review of the work developed by Pinto-Neto and Acacio de Barros is Pinto-Neto and Fabris (2013). For Valentini's works, see (Valentini 2007, 2010).

[7]The papers presenting Bohmian mechanics are Dürr et al. (1992a, b). Goldstein and Lebowitz (1995, 1208), Bohm (1952, 170).

Fig. 7.3 Goldstein, from Rutgers University, is one of the leaders of Bohmian mechanics. *Credits* Picture by Olival Freire

in all kinds of problems, law of large numbers type considerations, play a crucial role."[8] Thus for them, "Bohmian mechanics is a version of quantum mechanics for non relativistic particles in which the word 'particle' is to be understood literally: In Bohmian mechanics quantum particles have positions, always, and follow trajectories. These trajectories differ, however, from the classical Newtonian trajectories." Without referring to the quantum potential and the difficult problem of its physical interpretation, with this approach they derived the same results obtained both with standard quantum mechanics and with Bohm's original approach for nonrelativistic phenomena. As Goldstein put it in an acronym, their theory is simply the adoption of an OOEOW (Obvious Ontology Evolving the Obvious Way). One should note that when these physicists define what they understand to be a Bohmian theory, the preference for determinism disappears and they consider that "a Bohmian theory should be based upon a clear ontology," meaning by ontology "what the theory is fundamentally about." While for non-relativistic physics they adopt a particle ontology, they admit that they "have no idea what the appropriate ontology for relativistic

---

[8]Dürr et al. (1992a). Interview of Goldstein by Freire Junior on 2017 November 3, Niels Bohr Library and Archives, American Institute of Physics, College Park, MD USA.

physics actually is." As such, the commitment to a quantum ontology comes before an engagement with a causal pattern for physical theories, a position analogous to what had been adopted by Bohm and Hiley since the 1960s.[9]

Bohmian mechanics is useful for discussing quantum chaos and for this reason it has received acceptance beyond those interested in foundations of quantum mechanics. In the usual studies of chaotic systems, so widespread from the 1980s on, quantum chaos was limited to the measurement processes, as position and speed were not simultaneously well defined in quantum mechanics. Indeed, in classical chaos this chaoticity derived from the very dynamic laws. Dürr and colleagues have argued that if we adopt the assumptions of Bohmian mechanics, quantum chaos may arise in a similar manner to classical chaos.[10]

Dürr, Sheldon, Zanghi, and colleagues have been very active over the years battling for the diffusion of Bohmian mechanics. In particular they have criticized the long lasting adhesion of physicists to the standard approaches to quantum mechanics as just inertial adhesion to a received view. As Goldstein put it, "that's one of the difficulties getting people to be able to think clearly and judge clearly about issues in foundations of quantum mechanics, to overcome all of the misinformation that they get from their teachers." However, he clarified, "I'm not saying deliberate misinformation. The teachers are busy and the professors are busy. The scientists and physicists are busy. They have all kinds of problems they're working on. They naturally basically accept what has been accepted by the community. It's difficult to overcome that." They also took distance from Bohm's later approach, both from the revival of the quantum potential and from the ideas of wholeness and implicate order, as we have seen. One of their favorite mottoes to introduce Bohmian mechanics has been "Quantum Physics Without Quantum Philosophy," which is now the title of their jointly authored book. This motto is intended to criticize the putative need of "drastic conceptual innovations" for "a genuine explanation of quantum interference effects and quantum randomness." Attractive as this may be, it should not be taken at its face value. We should remember that taking as an ontology either particles or whatever else is also a philosophical stance. Leaving aside the over simplifications such mottoes may bring, Goldstein summarized his stance as follows[11]:

> By extending Bohmian mechanics I certainly do not mean having a theory involving a quantum potential. That's already clear from the beginning. I also do not mean a theory necessary in which the fundamental ontology, the primitive ontology, is a particle ontology. That's not what I mean by a Bohmian kind of theory. For me, a Bohmian theory is, first of all, a quantum theory. So the wave function plays a crucial role in the dynamics, but it plays a crucial role in the dynamics of something else. It's that something else which brings, makes the theory a physical theory. It brings it into contact with physics. Now what do you need to have contact with physics? You need to have objects, structures, clear structures in space and time. You need to have an ontology of what Bell called local beables. One of the obvious possibilities for local beables are particles with positions following world lines.

[9]Dürr et al. (1992a, b, 1996), Dürr et al. (2009).

[10]Dürr et al. (1992b).

[11]Interview of Goldstein by Freire Junior, op. cit. The book is Dürr et al. (2013).

The last strand of Bohm's scientific heritage is championed by Hiley, who continues to work on the research that he and Bohm had been carrying out in the last decade before Bohm's death. A short timeline of Bohm's thoughts on the interpretation of quantum mechanics may illustrate where Hiley gave continuity to his work with Bohm and where this strand departed from the previous one. Let us call the two previous strands by the names of some of their chief proponents. The first was that followed by Holland, and let us take his textbook as its reference, and the second led by Goldstein, and similarly let us take the Bohmian Mechanics' book as its reference.[12] As stated by Hiley, while recollecting his times at Birkbeck with Jeffrey Bub for a volume of *Foundations of Physics* honoring Bub, "one of the remarkable things about the sixties at Birkbeck was that, as far as I recall, we never discussed in any detail Bohm's original 1952 paper that forms the basis of what later became known as 'Bohmian mechanics'." Furthermore, according to Hiley, "Bohm himself was looking for a more fundamental structure of ideas upon which to base a new quantum theory. He was more interested in developing a process based philosophy in which new forms of order and structure would play a dominant role. It was, in fact, the beginning of what he eventually called the implicate order." Thus, both Holland and Goldstein departed from Bohm's early work. Holland extended it to include a random subquantum level while Goldstein rederived Bohm's conclusions in a simpler and cleaner way. Both based their work on Bohm's 1952 papers and extended them in order to obtain a better connection with quantum physics, including quantum field theories. They are, in this sense, the most direct heirs of Bohm's 1952 hidden variable interpretation of quantum mechanics. As we have seen, the evolution of Bohm's ideas concerning the interpretation of quantum mechanics was rather complex. In the 1980s, Bohm came back to his early hidden variable interpretation, changed his view about the quantum potential and tried to connect this early interpretation with the implicate order's ideas, as expressed in the textbook he and Hiley published in the early 1990s. As an illustration of their early joint work, in 1984 Bohm and Hiley generalized the mathematical structure of the Penrose twistor theory to a Clifford algebra. This work allowed "basic geometric forms and relationships to be expressed purely algebraically." Furthermore, Bohm and Hiley considered that "by means of an inner automorphism of this algebra, it is possible to regard these forms and relationships as emerging from a deeper pre-space," which authorized them to call this pre-space "an implicate order." In doing this, they were able to express basic geometric forms and relationships in pure algebraic language. This work, typical of mathematical physics, was intended to concretize their ideas on the interpretation of quantum mechanics. In their words, "by means of an inner automorphism of this algebra, it is possible to regard these forms and relationships as emerging from a deeper pre-space, which we are calling an implicate order." They expected such a work could open the way "for a new mode of description, that does not start from continuous space-time, but which allows this to emerge as a limiting case." In the following stage, Hiley departed from his previous work with Bohm, as expressed in *The Undivided Universe* textbook, and his later research tried to extend the insights concerning the mix of implicate

---

[12]Holland (1993), Dürr et al. (2013).

order and active information and the quest for algebraic structures able to underpin spacetime geometry and standard quantum mechanics (Fig. 7.4).[13]

In recent decades Hiley has kept up intense scientific activity along these lines he had been working on before Bohm's death. Indeed, Hiley's activities demonstrate that the compatibility between hidden-variables and wholeness and implicate order he and Bohm reached in the book *Undivided Universe* was not a satisfactory synthesis. It is beyond the scope of this book to provide a review of these activities, thus we are going to illustrate them through three examples. Hiley and Fernandes, from the University of Brasilia and his former doctoral student, have attempted to mathematically describe processes, which for Hiley should be considered the most basic ontological elements, without any initial reference to space and time. They reviewed earlier works where spatial properties may be abstracted from some specific mathematical structure, Clifford algebras in the case, and generalized these ideas to abstract a notion of time. For them, and following Bohm, "in the implicate order all moments of time were present together in what can be regarded as a timeless order. In spite

Fig. 7.4 Hiley was Bohm's collaborator for more than thirty years, at Birkbeck College, London. Together they developed the wholeness and implicate order approach and wrote the book The Undivided Universe. From left to right is Hiley, Vincenzo Monachello, Joel Morley and Peter Edmunds at University College, London, February 2016. On the lab benches there are the argon experiments to obtain weak measurements. *Credits* Photo by Freire

---

[13] Hiley (2009), Bohm and Hiley (1993, 1984). Other works along the same lines were Frescura and Hiley (1980a, b, 1984, 1987), and Bohm and Hiley (1981).

of this, the order of time could unfold through a series of explicate orders." Hiley and Fernandes thought the contribution of their work was that "what seemed like a vague general notion can now be given a more explicit mathematical form." Hiley inherited from Bohm's early hidden variables interpretation the major challenge of obtaining a fully relativistic treatment in order to match the level attained by standard quantum mechanics with the Dirac equation. Working on this algebraic approach he eventually showed that the dynamics of particles ruled by the Schrödinger, Pauli and Dirac equations can be described in a hierarchy of Clifford algebras. In a series of papers with Robert Edward Callaghan, from Birkbeck College, they applied Clifford algebras to the Bohm approach, as outlined in Bohm and Hiley's 1993 book, to provide "for the first time, an elegant, unified approach to the Bohm model of the Schrödinger, Pauli and Dirac particles, in which we no longer have to appeal to any analogy to classical mechanics to motivate the approach as was done by Bohm in his original paper." According to Hiley and Callaghan, the algebraic approach adopted does not suffer from the difficulties of earlier attempts because it does not require the use of wave functions. This is because "the information normally encoded in the wave function is already contained within the algebra itself, [...] and there is no need to separate operator from operand." This allowed them to "have a natural a way to embed the physical properties of these particles in a set of successive approximations, relativistic particle with spin, non-relativistic particle with spin, non-relativistic particle without spin, in one mathematical structure, a mathematical structure that naturally arises from the geometric structure of space-time."[14]

In recent years Basil Hiley has been engaged in interpreting the results obtained with "trajectories" of quantum particles. The subject has gained some traction following the notion of "weak values" of the momentum operators introduced by Aharonov et al. (1988) and the work by Sacha Kocsis and co-workers, who showed, in 2011, the average trajectories of single photons in a two-slit device (Kocsis et al. 2011). Joining efforts with Robert Flack, from University College London, Hiley was able to arrive at a number of startling results. They showed the relationship between weak values, Feynman's transition probability amplitudes, and Bohm's approach. In the boldest conclusions, they argued that the paths obtained by Kocsis and co-workers were not photon trajectories because "photons cannot be treated as particles that satisfy the Schrödinger equation. They have zero rest mass and are excitations of the electromagnetic field." However, these paths may be interpreted as the "flow lines from an average made over many individual input photons. Thus, the so-called 'photon' flow-lines are constructed *statistically* from an ensemble of individual events." This conclusion allowed them to rectify a view, shared by contemporary supporters of Bohmian mechanics and by Hiley himself and colleagues in the past. As they argued "the basic assumption made in Bohmian mechanics, namely, that each particle follows one of the ensemble of 'trajectories' calculated by Philippidis et al. [in 1979]

---

[14]Hiley and Fernandes (1997), Hiley and Callaghan (2012). For a list of Basil Hiley's works, till 2013, see the journal *Foundations of Physics*, Volume 43(4), 415–423, April 2013. This issue includes a Festschrift for him, edited by Chris Dewdney, Paavo Pylkkänen, and Harald Atmanspacher. Bohm and Hiley (1993).

from P_B (x, t) cannot be maintained. Rather the trajectories should be interpreted as a statistical average of the momentum flow of a basic underlying stochastic process." They expect that current experiments being performed with atoms instead of photons may shed more light on the subject.[15]

Hiley's continued work on the foundations of quantum mechanics, particularly that following Bohm's passing, was well synthesized by Atmanspacher in these words[16]:

> The idea to find a mathematically formalizable background theory behind Hilbert space quantum mechanics has kept occupying Basil in the twenty years from 1990 to the present. In these two decades, he has made significant steps beyond the joint work with Bohm, suggested new ideas, obtained new results, and achieved new insights. The basic move is quite straightforward: identify representation-free structures whose representation (for instance in a Hilbert space) reinstantiates the known aspects of quantum mechanics.

How we may interpret this diversity of followers of Bohm's intellectual, particularly scientific, legacy? First, it should be seen as a compliment as the very existence of such a diversity reveals the fecundity of Bohm's thoughts. Ultimately the controversy among Goldstein's, Holland's, and Hiley's ideas, to keep with these leader scientists, mirrors, at least to a certain extent, the wider controversy over the interpretation of quantum mechanics. Thus an evaluation of this diversity of intellectual followers requires an consideration of the wider context of the long controversy over the interpretation of quantum mechanics, the controversy which was the butter and bread for Bohm during four decades.

Intellectual controversies are more common in scholarly fields such as philosophy, humanities at large, and social sciences. However, it is not strange to the history of the natural sciences. It may be less known among wider audiences but it is well set particularly among historians of science. Every time we deal with issues concerning the foundations or the interpretations of scientific theories, or, still, with the reception of such theories, we stumble upon analogous diversities. Our case may be illustrated with past episodes such as the reception of Newton's mechanics and gravity; the reception and later developments of Darwin's theory of natural selection; the reception of Einstein's principle of relativity; Mach's criticisms towards Newton's mechanics; the reception and interpretation of Wegener's continental drift theory; controversies over geological eras; debates over action at a distance and action by continuity in electromagnetism; debates on 20th century cosmological models; the long controversy over the foundations of second law of thermodynamics; and debates during the coalescence of quantum chemistry as a new subdisciplinary field. The list is too long for any comprehensive account in this book. Similarly, the number of historical studies on these topics is too extensive to summarize here; thus I just cite a few references for readers interested in following these episodes further.[17]

---

[15]Flack and Hiley (2018).

[16]Atmanspacher (2013).

[17]A very short list of historical studies of these controversies are: on the reception of Newton's mechanics and gravity, Blay (1995, 2002); on the reception and later developments of Darwin's natural selection, Bowler (2003); on the reception of Einstein's principle of relativity, Glick (1987)

Indeed, controversies in science are more common than scientists themselves acknowledge. To illustrate my point, I use images of science built by some 20th century philosophers of science. I would say that the real image of scientific activities stands closer to Lakatos and Feyerabend's views, with their proliferation of controversies, than to Thomas Kuhn's view with the dominance of "normal science." In Kuhn's views, controversies are sent to science's backstage and only appear in exceptional times, those of scientific revolutions.[18]

Against this backdrop Bohm's life and works are less unique than they appear at the first view. Their uniqueness lies however, in the intensity and duration of these controversies. He was a key protagonist of the intense and lasting controversy over the interpretation of quantum physics and his legacy has triggered other chapters of this very controversy. Furthermore, if we now understand the quantum puzzles better, or, at least, we understand better the challenges we face to deal with them, these achievements bear the imprints of Bohm's long quest to understand quantum physics. He was a great quantum dissident and his dissidence has contributed to a better understanding of the quantum world.

References

Aharonov, Y., Albert, D.Z., Vaidman, L.: How the result of a measurement of a component of the spin of a spin ½ particle can turn out to be 100. Phys. Rev. Lett. **60**(14), 1351–1354 (1988)

Atmanspacher, H.: Appreciating a Hiley respected colleague. Found. Phys. **43**, 412–414 (2013)

Barbour, J.B., Pfister, H. (eds.): Mach's Principle: From Newton's Bucket to Quantum Gravity. Birkhäuser, Boston (1995)

Bell, J.S.: On the impossible pilot wave. Found. Phys. **12**(10), 989–999 (1982)

Bell, J.S.: Beables for quantum field theory. In: Hiley, B.J., Peat, F.D. (eds.) Quantum Implications: Essays in Honour of David Bohm, pp. 227–234. Routledge & Kegan, London (1987)

Besson, V.: L'interprétation causale de la mécanique quantique: biographie d'un programme de recherche minoritaire (1951–1964). Ph.D. Dissertation, Université Claude Bernard Lyon 1 and Universidade Federal da Bahia (2018)

Blay, M.: Les Principia de Newton. Presses universitaires de France, Paris (1995)

Blay, M.: La science du mouvement: de Galilée à Lagrange. Belin, Paris (2002)

Bohm, D.: A suggested interpretation of the quantum theory in terms of hidden variables—I & II. Phys. Rev. **85**(2), 166–179 and 180–193 (1952)

Bohm, D.: Wholeness and the Implicate Order. Routledge & Kegan Paul, London (1980)

---

and Paty (1993); on Mach's criticisms towards Newton's mechanics, Barbour and Pfister (1995); on the reception and interpretation of Wegener's continental drift theory, Oreskes (1988) and Frankel (2012); on controversies over geological eras, Rudwick (1985); on debates over the action at a distance and action by continuity in electromagnetism, Darrigol (2000); on debates on the 20th century cosmological models, Kragh (1996); for a short presentation of the issues concerning the long controversy over the foundations of the second law of thermodynamics, see Mitchell (2009), particularly Chap. 3, pp. 40–55, and the references therein; and on debates during the coalescence of quantum chemistry as a new subdisciplinary field, see Gavroglu and Simões (2012).

[18]Lakatos (1978), Feyerabend (1993), Kuhn (1962).

Bohm, D., Biederman, C.J., Pylkkänen, P.: Bohm-Biederman Correspondence. Routledge, London (1999)

Bohm, D., Hiley, B.J.: On a quantum algebraic approach to generalised phase space. Found. Phys. **11**, 179–203 (1981)

Bohm, D., Hiley, B.J.: Generalisation of the twistor to Clifford algebras as a basis for geometry. Revista Brasileira de Física. Volume Especial (Os 70 anos de Mario Schönberg) 1–26 (1984)

Bohm, D., Hiley, B.J.: The Undivided Universe: An Ontological Interpretation of Quantum Theory. Routledge, London (1993)

Bowler, P.J.: Evolution: The History of an Idea. University of California Press, Berkeley (2003)

Cross, A.: The crisis in physics: dialectical materialism and quantum theory. Soc. Stud. Sci. **21**, 735–759 (1991)

Cushing, J.: Quantum Mechanics—Historical Contingency and the Copenhagen Hegemony. The University of Chicago Press, Chicago (1994)

Darrigol, O.: Electrodynamics from Ampère to Einstein. Oxford University Press, New York (2000)

Dürr, D., Goldstein, S., Zanghi, N.: Quantum equilibrium and the origin of absolute uncertainty. J. Stat. Phys. **67**(5/6), 843–907 (1992a)

Dürr, D., Goldstein, S., Zanghi, N.: Quantum chaos, classical randomness, and Bohmian mechanics. J. Stat. Phys. **68**(1–2), 259–270 (1992b)

Dürr, D., Goldstein, S., Zanghi, N.: Bohmian mechanics at the foundation of quantum mechanics. In: Cushing, J.T., Fine, A., Goldstein, S. (eds.) Bohmian Mechanics and Quantum Theory: An Appraisal, pp. 21–44. Kluwer, Dordrecht (1996)

Dürr, D., Goldstein, S., Zanghi, N.: Quantum Physics Without Quantum Philosophy. Springer, Berlin (2013)

Dürr, D., Goldstein, S., Tumulka, R., Zanghi, N.: Bohmian mechanics. In: Greenberger, D., Hentsche, K., Weinert, F. (eds.) Compendium of Quantum Physics—Concepts, Experiments, History and Philosophy, pp. 47–55. Springer, Berlin (2009)

Feyerabend, P.: Against Method: Outline of an Anarchistic Theory of Knowledge. Verso, London (1993)

Flack, R., Hiley, B.J.: Feynman paths and weak values. Entropy **20**, 367–378 (2018)

Forstner, C.: Quantenmechanik im Kalten Krieg: David Bohm und Richard Feynman. GNT-Verlag, Stuttgart (2007)

Forstner, C.: The early history of David Bohm's quantum mechanics through the perspective of Ludwik Fleck's thought-collectives. Minerva **46**(2), 215–229 (2008)

Frankel, H.R.: The Continental Drift Controversy, (4 vols.). Cambridge University Press, New York (2012)

Freire Junior, O.: David Bohm e a controvérsia dos quanta. Centro de Lógica, Epistemologia e História da Ciência, Campinas (Brazil) (1999)

Freire Junior, O.: Science and exile: David Bohm, the cold war, and a new interpretation of quantum mechanics. Hist. Stud. Phys. Biol. Sci. **36**(1), 1–34 (2005)

Freire Junior, O.: Continuity and change: charting David Bohm's evolving ideas on quantum mechanics. In: Krause, D., Videira, A. (eds.) Brazilian Studies in Philosophy and History of Science, pp. 291–299. Springer, Heidelberg (2011)

Freire Junior, O.: The Quantum Dissidents—Rebuilding the Foundations of Quantum Mechanics 1950–1990. Springer, Berlin (2015)

Frescura, F.A.M., Hiley, B.J.: The implicate order, algebras and the spinor. Found. Phys. **10**, 7–31 (1980a)

Frescura, F.A.M., Hiley, B.J.: The algebraisation of quantum mechanics and the implicate order. Found. Phys. **10**, 705–722 (1980b)

Frescura, F.A.M., Hiley, B.J.: Algebras, quantum theory and pre-space. Revista Brasileira de Física **14**, 49–86. Volume Especial (Os 70 anos de Mario Schönberg) (1984)

Frescura, F.A.M., Hiley, B.J.: Some spinor implications unfolded. In: Hiley, B.J., Peat, F.D. (eds.) Quantum Implications: Essays in Honour of David Bohm, pp. 278–288. Routledge, London (1987)

Gavroglu, L., Simões, A.: Neither Physics nor Chemistry. A History of Quantum Chemistry. The MIT Press, Cambridge, MA (2012)

Glick, T.F. (ed.): The Comparative Reception of Relativity. Reidel, Dordrecht (1987)

Goldstein, S.: Obituaries: David Joseph Bohm. Physics Today 47, 8, p. 72 (1994)

Goldstein, S., Lebowitz, J.: Quantum mechanics. In: Stroke, H.H. (ed.) The Physical Review—The First Hundred Years: A Selection of Seminal Papers and Commentaries, pp. 1205–1214. American Institute of Physics, New York (1995)

Hiley, B.J.: David Joseph Bohm. Biogr. Mem. Fellows R. Soc. 43, 106–131 (1997)

Hiley, B.J.: On the relationship between the Wigner-Moyal and Bohm approaches to quantum mechanics: a step to a more general theory? Found. Phys. 40, 356–367 (2009)

Hiley, B.J., Callaghan, R.E.: Clifford algebras and the Dirac-Bohm Quantum Hamilton-Jacobi equation. Found. Phys. 42, 192–208 (2012)

Hiley, B., Fernandes, M.C.B.: Process and time. In: Atmanspacher, H., Ruhnau, E. (eds.) (Org.) Time, Temporality, Now, pp. 365–383. Springer, New York (1997)

Holland, P.R.: The Quantum Theory of Motion: An Account of the de Broglie-Bohm Causal Interpretation of Quantum Mechanics. Cambridge University Press, Cambridge (1993)

Holland, P.: Jean-Pierre Vigier at seventy-five: La Lutte continue. Found. Phys. 25(1), 1–4 (1995)

Horgan, J.: Last words of a quantum heretic. New Scientist, 27 Feb 1993, pp. 38–42 (1993)

Kocsis, S., Braverman, B., Ravets, S., Stevens, M.J., Mirin, R.P., Shalm, L.K., Steinberg, A.M.: Observing the average trajectories of single photons in a two-slit interferometer. Science 332, 1170–1173 (2011)

Kojevnikov, A.: David Bohm and collective movement. Hist. Stud. Phys. Biol. Sci. 33, 161–192 (2002)

Kragh, H.: Cosmology and controversy: the historical development of two theories of the universe. Princeton University Press, Princeton, N.J. (1996)

Kuhn, T.S.: The Structure of Scientific Revolutions. University of Chicago Press, Chicago (1962)

Lakatos, I.: The Methodology of Scientific Research Programmes. Cambridge University Press, New York (1978)

Mitchell, M.: Complexity—A Guided Tour. Oxford University Press, New York (2009)

Mullet, S.K.: Little man: four junior physicists and the red scare experience. Ph.D. Dissertation, Harvard University (2008)

Olwell, R.: Physical isolation and marginalization in physics—David Bohm's cold war exile. ISIS 90, 738–756 (1999)

Oreskes, N.: The rejection of continental drift. Hist. Stud. Phys. Biol. Sci. 18, 311–348 (1988)

Pace, E.: David J. Bohm, 74, Physicist and writer on quantum theory. The New York Times, 29 Oct 1992, p. 16 (1992)

Paty, M.: Einstein Philosophe: La Physique Comme Pratique Philosophique. Presses universitaires de France, Paris (1993)

Peat, F.D.: Infinite Potential: The Life and Times of David Bohm. Addison Wesley, Reading, MA (1997)

Pinch, T.: What does a proof do if it does not prove? A study of the social conditions and metaphysical divisions leading to David Bohm and John von Neumann failing to communicate in quantum physics. In: Mendelsohn, E., Weingart, P., Whitley, R. (eds.) The Social Production of Scientific Knowledge, pp. 171–216. Reidel, Dordrecht (1977)

Pinto-Neto, N., Fabris, J.C.: Quantum cosmology from the de Broglie-Bohm perspective. Class. Quantum Gravity 30, 143001 (2013)

Pylkkänen, P.: Mind, Matter and The Implicate Order. Springer, Berlin (2007)

Redhead, M.L.G.: David Bohm (1917–1992) [Obituary]. Nature 360, 12 Nov 1992, p. 107 (1992)

Rudwick, M.J.S.: The Great Devonian Controversy: The Shaping of Scientific Knowledge among
    Gentlemanly Specialists. University of Chicago Press, Chicago (1985)
Stroke, H.H. (ed.): The Physical Review—The First Hundred Years: A Selection of Seminal Papers
    and Commentaries. American Institute of Physics, New York (1995)
Talbot, C. (ed.): David Bohm: Causality and Chance, Letters to Three Women. Springer, Berlin
    (2017)
Valentini, A.: Astrophysical and cosmological tests of quantum theory. J. Phys. A Math. Theor.
    40(12), 3285–3303 (2007)
Valentini, A.: Inflationary cosmology as a probe of primordial quantum mechanics. Phys. Rev. D
    82(6) (2010)

Chapter 8
The Legacy of David Bohm in Physics—An Essay in Scientometry

8.1 Introduction

In the previous chapters I have analyzed the ensemble of David Bohm's scientific works, including their concepts, their mathematical structure, their interaction with experiments, their influence among physicists, and his recognition among this professional community. Now I take a slightly different approach to Bohm's works. In this chapter I use scientometry, particularly the number of citations of his papers by other papers and the evolution of these citations over time, to assess the influence, as measured by this quantitative marker, of the works of David Bohm in the physics community.

In this final chapter I would like to address the following question: In order to assess Bohm's legacy to physics I shall ask what Bohm's persona inside the expert circle of physicists was? However, I ask this question now through the lens of scientometry, particularly the number of citations of his papers, taking into consideration its intrinsic limitations. Thus I add to these data some qualitative information concerning how these papers were seen by other physicists. Bohm was a long-standing critic of the standard interpretation of quantum mechanics, thus a quantum dissident, and devoted much of his professional life to an alternative interpretation of quantum theory, the causal interpretation.[1] He suggested it at a time, the early 1950s, when research on the foundations of quantum mechanics was poorly viewed among physicists. Most of them considered foundational issues already solved by the founding fathers of this physical theory. As time went by interest in foundational issues widened, indeed it became a regular field for physical research, and Bohm's works went on to be considered more positively, or at least with more tolerance. I'd like to go beyond this framework here and instead of considering Bohm's prestige as a kind

---

An abridged version of this chapter appeared in Michael R. Matthews (ed). Mario Bunge: Centenary Festschrift, Springer, 2019.

[1] On the quantum dissidents, see Freire Junior (2015, p. 2).

© Springer Nature Switzerland AG 2019
O. Freire Junior, *David Bohm*, Springer Biographies,
https://doi.org/10.1007/978-3-030-22715-9_8

of mirror of the prestige of the research in foundations, to look in a more detailed manner at the ensemble of Bohm's achievements and how these achievements were received by the public, in this case, other physicists.

In the second section of this chapter I discuss the advantages and the limitations of the use of scientometry to assess achievements and failures in science and in the history of science. I go on to suggest and justify the adoption of a certain metric for the identification of Bohm's most cited papers and present a qualitative overview of Bohm's achievements. In the third section I present and analyze the list of Bohm's most cited scientific papers. In the following sections, fourth to sixth, I successively analyze three groups of Bohm's most cited papers, namely, Plasma and Collective Variables, Reinterpretation of Quantum Mechanics, and Quantum Effects and Experiments. The seventh section is dedicated to considering Bohm's persona through some qualitative comments from his contemporary fellow physicists. Finally, in the concluding remarks I resume my assessment of the influence of Bohm's legacy to physics bringing together both quantitative and qualitative considerations.

8.2 Scientometry in Science and History of Science

Scientometry has become a widely used resource to assess the production of science. It has been used to assess science's strength and weakness in the case of countries, institutions, disciplines, and research teams as well as individual scientists. Number of papers, number of citations, number of collaborators, individual h-index, and journal impact factor are now part of the lexicon of scientific and educational managers and individual scientists. They have also been used in the studies on science, particularly social science studies and history of science. However, the careless use of scientometry, i.e. without taking into consideration its limitations, can hamper a better understanding of science as it is indeed produced. A recent caveat of this was the so-called Leiden Manifesto, which was signed by Diana Hicks, Paul Wouters, Ludo Waltman, Sarah de Rijcke, and Ismael Rafols. According to these authors, "data are increasingly used to govern science. Research evaluations that were once bespoke and performed by peers are now routine and reliant on metrics." The problem "is that evaluation is now led by the data rather than by judgement. Metrics have proliferated: usually well intentioned, not always well informed, often ill applied." They go on to conclude, "we risk damaging the system with the very tools designed to improve it, as evaluation is increasingly implemented by organizations without knowledge of, or advice on, good practice and interpretation." Among the risks, the authors cite the different regimes of citations across disciplines, thus making nonsensical comparisons of raw number of citations of scientists from distinct disciplines. They noted that "citation rates vary by field: top-ranked journals in mathematics have impact factors of around 3; top-ranked journals in cell biology have impact factors of about 30." Thus it comes as no surprise that if scientometrics is to be used in history of science, this should be carefully done and never used as a sole resource. This was the conclusion of historian Helge Kragh, while analyzing scientometric historiography: "it should be obvious that scientometrics cannot, in any circumstances, stand alone," and "if scientometry is used with care and in combination with other methods, it can play an important part, especially in the study of modern science." If this was the general recommendations, Kragh was rather more specific about the use of counting citations in history of science: "there is no doubt that in many cases frequency of citation does reflect impact, but the measure should not be credited with any particularly objective or reliable status compared with evaluations based on qualitative assessment." This is the reason I am using scientometry at the end of our journey throughout Bohm's works and life, as a supplementary resource to assess the influence of his scientific work.[2]

Having decided to use the number of citations as a marker of influence in a certain scientific community, I had to find a term of comparison for papers inside physics, particularly theoretical physics, in the second half of the 20th and early 21st centuries, in order to assess which of Bohm's papers could be considered highly influential. I took the number of 300 citations per paper as evidence of highly influential papers in physics. I also commented on the remainder in the list of Bohm's papers obtaining beyond of 100 citations. The figure of 300 citations is a guess, just a guess, but it is a guess informed by a certain number of results from scientometry. For instance, when I took data from the number of US article output and citations of US articles, in the field of physics, for the years 1997, 1999, and 2000, I obtained an average of 7.1 citations per article. Sid Redner, considering only citations in *Physical Review* of papers published in this journal, which is a top journal in physics, particularly after WWII, concluded that "nearly 70% of all PR articles have been cited fewer than 10 times" and that "the average number of citations is 8.8." Thus the average number obtained from these two distinct sources are not discrepant; furthermore, these numbers match the physicists' shared tacit perception that an article should receive more than 10 citations to be considered known.[3]

However, to obtain more trustworthy numbers we need normalized indicators as, according to Hicks and co-authors, "normalized indicators are required, and the most robust normalization method is based on percentiles: each paper is weighted on the basis of the percentile to which it belongs in the citation distribution of its field." Thus we need some indications from the number of citations of highly cited papers in physics. Gregory Patience and co-workers compiled a list of bibliometric indicators from the Web of Science for 236 science categories and citation rates of the 500 most cited articles of each category, and built a number of different indicators, one of them being the average number of citations of these highly cited papers. For applied physics, they found 311; for condensed matter physics, 296; for astronomy and astrophysics, 197; for atomic and molecular physics, 127; for particles and fields physics, 126; for mathematical physics, 61; fluids and plasmas physics, 55; spectroscopy, 53; and nuclear physics, 82. Some of these categories, for instance

---

[2](Hicks et al. 2015). Similar conclusions about citation patterns in different scholar field were reached by Podlubny (2005). Kragh (1987, 192–196). On the use of citation data in history of science, see also Freitas and Freire (2003).

[3]Data on US articles came from http://www.nsf.org; Redner (2005).

condensed matter, were not considered as such when David Bohm began his scientific career, while some of the methods he developed with Gross and Pines are now part of the condensed matter field. Keeping this in mind we may consider Bohm's works to fit in categories such as particles and fields, mathematical physics, and plasmas. We also took a look at Spires, the Stanford's database for high energy physics preprints. They suggest the following classification: "Unknown papers (0); Less known papers (1–9); Known papers (10–49), Well-known papers (50–99), Famous papers (100–499 cites), and Renowned papers (500+ cites)."[4] In conclusion, taking 300 citations per paper as an indication of a highly influential paper in physics is a tentative number in consonance with studies in scientometrics.

Before presenting and analyzing Bohm's citation data, let us summarize his achievements. His Ph.D. dissertation, on nucleon scattering, was considered relevant enough to be immediately classified in the context of WWII. Still at Berkeley he worked on the synchrocyclotron and on plasma. It can be said that his earliest work on plasmas and his later work, at Princeton, on collective coordinates made his reputation as physicist. However, his most original and heterodox contribution to physics may have been the elaboration of the hidden variables interpretation of quantum physics, published in 1952. This departed considerably from the standard theory in its conceptual and philosophical assumptions but still arrived at the same predictions, thus opening the way for alternative interpretations of quantum mechanics. In the late 1950s, Yakir Aharonov and Bohm's seminal paper contributed to our understanding of the role of phases and electromagnetic potentials in the quantum domain. Later, Bohm put the causal interpretation aside and turned his attention to the highly mathematical approach called implicate order, working with Basil Hiley, looking for the most basic algebraic structures from which quantum theories could emerge. In fact, while abandoning the quest for the recovery of a deterministic view of quantum phenomena, Bohm remained committed to a realistic view of physical theories. In the late 1970s, two students, Chris Dewdney and Chris Philippidis surprised Bohm and Hiley by building diagrams of the paths and potentials from Bohm's early interpretation using computers. Bohm then came back to his early hidden variable interpretation. At that time, interest in the foundations of quantum mechanics was being revived and Bohm spent his last years trying to reconcile his own different approaches to the quantum. This attempt is expressed in Bohm and Hiley's textbook, *The Undivided Universe*, published posthumously. Bohm's work generated several strands. They include Hiley's algebraic work; Bohmian mechanics, led by Detlef Dürr, Sheldon Goldstein, and Nino Zanghi; and attempts to use and test Bohm's ideas in the domains of cosmology.[5]

---

[4]Hicks et al. (2015). Patience et al. (2017). The Spires information came from http://www.slac.stanford.edu/spires/; these number were discussed in Freire Junior (2015, 585).

[5](Aharonov and Bohm 1959; Philippidis et al. 1979; Bohm and Hiley 1993).

8.3 Bohm's Most Cited Scientific Papers

Nowadays David Bohm has ten papers with more than 300 citations. They are listed in the table in Fig. 8.1.[6] We also have the number of papers in which the term "Bohm diffusion" appears as a topic. This brings the total to 345 papers. We will deal with the case of "Bohm diffusion" references in the next section. For the moment, we shall focus on the number of papers, ten, with over 300 citations. That is not bad for any scientist, in physics at least. Figure 8.1 contains a list of these papers in the order of number of citations.[7] The ensemble of these papers reveals some patterns in Bohm's contributions to physics which will be further analyzed later. All these ten papers may be arranged in three groups: collective variables (plasma and metals), reinterpretation of quantum mechanics, and quantum effects and experiments. In addition, all of them were published between 1949 and 1959. For the moment, however, I would like to comment that even if we reduce our search for papers to above 100 citations, there will be no changes in the patterns we are going to analyze. Indeed, with more than 100 citations and fewer than 300 the following papers are included: Aharonov and Bohm (1961a, b), Bohm and Bub (1966), Bohm et al. (1987), Bohm (1953) and Bohm and Weinstein (1948).[8] Except for the paper by Bohm and Weinstein, published in 1948, all these papers fall in the field of interpretation of quantum theory and quantum effects and experiments, or still, foundations of quantum mechanics in the case of the paper Aharonov and Bohm (1961a). Except for the 1987 paper co-authored by Bohm, Hiley and Kaloyerou, they roughly cover the period between the late 1940s and the early 1960s.

Let us now come back to the ten papers with above 300 citations. They were published, as we have noted, between 1949 and 1959. As Bohm got his Ph.D. during World War II and was active in physics till his death in 1992, our first conclusion is that Bohm's most influential stage in physics came a little after his Ph.D., lasted for ten years or so and then receded with less influential works.

These data allow us to deal with two thorny issues which appear when one reflects on Bohm's public image. First, Bohm is widely acknowledged for his dialogue with the Indian thinker Jiddu Khrisnamurti, and these dialogues have influenced some cultural circles. However, evidence of the influence of these exchanges on Bohm's more influential papers is scant. The reason for this is related to the fact that these dialogues began, as we have seen in Chaps. 5 and 6, in the early 1960s. Their influence on Bohm appeared in the centrality of the idea of wholeness, which was related to Krishnamurti's ideas about the roles of the observer and the observed as a whole. Before continuing, I would like to recall that the ideas of the Indian thinker were not alone in influencing Bohm on adopting the idea of wholeness, as we have seen in

---

[6]Aharonov and Bohm (1959), Bohm (1952a, b), Bohm and Gross (1949a, b), Pines and Bohm (1952), Bohm and Pines (1951, 1953), Bohm and Vigier (1954), Bohm and Aharonov (1957).

[7]These data were obtained from the Web of Science—Core Collection, from 1945 to 2018, accessed on 07 December 2018.

[8]Aharonov and Bohm (1961a, b), Bohm and Bub (1966), Bohm et al. (1987), Bohm (1953), and Bohm and Weinstein (1948).

Bohm's papers with over 300 citations			
Paper	*Authors*	*Year*	*Citations*
1 - Significance of electromagnetic potentials in the quantum theory	Aharonov & Bohm	1959	4,311
2 - A suggested interpretation of the quantum theory in terms of hidden variables - 1	Bohm	1952	2,946
3 - A suggested interpretation of the quantum theory in terms of hidden variables - 2	Bohm	1952	1,379
4 - A collective description of electron interactions - 3. Coulomb interactions in a degenerate electron gas	Bohm & Pines	1953	1,365
5 - A collective description of electron interactions - 2. Collective vs individual particle aspects of the interactions	Pines & Bohm	1952	796
6 - Theory of plasma oscillations - A. Origin of medium-like behavior	Bohm & Gross	1949	607
7 - A collective description of electron interactions - 1. Magnetic interactions	Bohm & Pines	1951	414
8 - Model of the causal interpretation of quantum theory in terms of a fluid with irregular fluctuations	Bohm & Vigier	1954	383
9 - Theory of plasma oscillations - B. Excitation and damping of oscillations	Bohm & Gross	1949	351
10 - Discussion of experimental proof for the paradox of Einstein, Rosen, and Podolsky	Bohm & Aharonov	1957	334

Fig. 8.1 List of the ten scientific papers by David Bohm which obtained more than 300 citations. Data from Web of Science—Core Collection, ranging from 1945 to 2018, accessed on 07 December 2018

the previous chapter. Wholeness and order coalesced as a research program in early 1970 and was passed to the wider audience in Bohm's book *Wholeness and Implicate Order*, published in 1980. None of the ten most cited of Bohm's paper fits into this chronological and conceptual framework. If we move to the papers with over than 100 citations, only the paper co-authored by Bohm, Hiley, and Kaloyerou, published in 1987, may reflect some interaction with Krishnamurti's ideas as the first part of this paper, signed only by Bohm and Hiley, made a slight reference to the idea of wholeness. Thus, if these dialogues had any influence on Bohm's ideas, and it had, as we have seen in the previous Chaps. 5 and 6, they did not have any bearing on the Bohm's most cited papers.

Second, Bohm worked for years on a philosophical program he called implicate order. In scientific terms this program led to attempts to obtain quantum mechanics and geometry departing from some primary algebraic structures. In this program Bohm worked collaboratively with Basil Hiley, his long standing colleague at Birkbeck College in London. Interesting as this program might have been, and it was in my opinion, for the future development of physics, it has yet to bear fruit in terms of its influence when citations are concerned. Indeed none of Bohm's papers resulting from the implicate order program appear among his ten most cited papers. Again, among the papers with over than 100 citations, only the paper co-authored by Bohm, Hiley, and Kaloyerou, published in 1987, is an expression of this research program. Thus neither the dialogue with Krishnamurti nor the implicate order program were as influential among his fellow physicists as Bohm's other works, that is, his most cited papers. We arrive at this conclusion constrained by our decision to look for Bohm's legacy to physics through the data from the number of citations of his papers, that is scientometry, a traditional but not unproblematic source for historians of science.

Now let us rearrange these data and group these ten papers according the subject covered. In the first group are the five papers related to Plasma and Methods for Condensed Matter, which we call Collective Variables—Plasma and Metals. They were produced by Bohm and his doctoral students Eugene Gross and David Pines and were published from 1949 to 1953. They are the papers number 4, 5, 6, 7, and 9. The second group concerns the Reinterpretation of Quantum Mechanics, papers 2, 3 and 8. The two first were produced by Bohm alone and the third through collaboration with Jean-Pierre Vigier. They were published between 1952 and 1954. The last group concerns Quantum Effects and Experiments. The papers resulted from Bohm's work with his doctoral student Yakir Aharonov. They were published from 1957 to 1959 and are the papers number 1 and 10.

8.4 Collective Variables—Plasma and Metals

Let us now take the five papers related to Collective Variables. Bohm worked on plasma with Eugene Gross, resuming work he had begun during World War II, in particular with Erich Burhop and Harrie Massey, who were part of the team of British physicists who went to Berkeley to science work related to the American atomic project. He went on to extend this approach to metals with David Pines. In addition to these five papers, two with Gross and three with Pines, there was another paper, by Pines alone. These were the main papers in which the collective variables approach was developed. The historian of physics Alexei Kojevnikov has studied Bohm's early approach to plasma and his evolution to the use of "collective variables" to deal with the phenomena of plasma and electrons in metals as well as the philosophical assumptions behind Bohm's approach. Kojevnikov also made an analysis of the wide range influence of this approach among condensed matter physicists and on the particular forms this approach received through its later circulation.[9] These papers have been steadily cited since their publication. This means that they have always been influential and the number of citations did not change dramatically either due to internal or external changes in the field. In Figs. 8.2, 8.3, 8.4, 8.5 and 8.6, we take the evolution of number of citations over the years for each of these papers just to

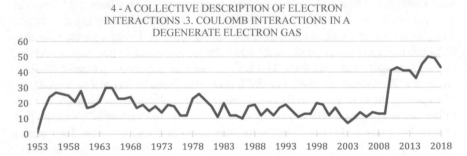

Fig. 8.2 Graph of the number of citations over the years for the paper "A collective description of electron interactions-3. Coulomb interactions in a degenerate electron gas," by Bohm and Pines (1953). Data from Web of Science—Core Collection, ranging from 1945 to 2018, accessed on 07 December 2018

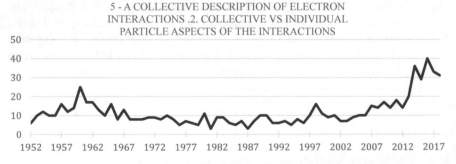

Fig. 8.3 Graph of the number of citations over the years for the paper "A collective description of electron interactions-2. Collective vs individual particle aspects of the interactions," by Pines and Bohm (1952). Data from Web of Science—Core Collection, ranging from 1945 to 2018, accessed on 07 December 2018

---

[9](Pines 1953), Kojevnikov (2002).

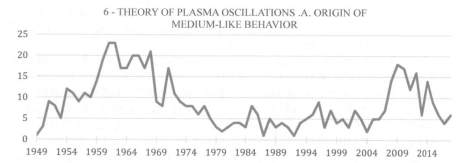

Fig. 8.4 Graph of the number of citations over the years for the paper "Theory of plasma oscillations-A. Origin of medium-like behavior," by Bohm and Gross (1949a). Data from Web of Science—Core Collection, ranging from 1945 to 2018, accessed on 07 December 2018

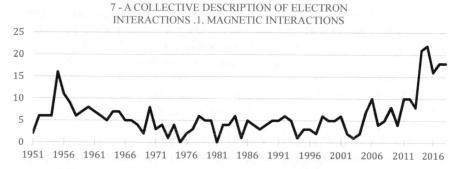

Fig. 8.5 Graph of the number of citations over the years for the paper "A collective description of electron interactions-1. Magnetic interactions," by Bohm and Pines (1951). Data from Web of Science—Core Collection, ranging from 1945 to 2018, accessed on 07 December 2018

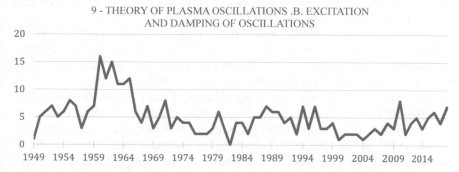

Fig. 8.6 Graph of the number of citations over the years for the paper "Theory of plasma oscillations-B. Excitation and damping of oscillations," by Bohm and Gross (1949b). Data from Web of Science—Core Collection, ranging from 1945 to 2018, accessed on 07 December 2018

illustrate their steady flow of citations. Some of the papers received slightly more attention just after their publication and also in recent years but overall they attracted a steady flow of attention among physicists.

Collective description, initially intended to deal with plasma and electrons in metals, was further extended, particularly by David Pines, to other systems including superconductivity and nuclear physics. This approach directly contributed to the wide class of problems in physics related to many-body systems, such as the cited cases of superconductivity and nuclear physics. Evidence of the circulation of these ideas is reported in Ben Mottelson's Nobel Prize speech, when he addressed the evolution of the models of nuclear physics and recalled: "it was a fortunate circumstance for us [Mottelson and Aage Bohr] that David Pines spent a period of several months in Copenhagen in the summer of 1957, during which he introduced us to the exciting new developments in the theory of superconductivity."[10] This interaction had been earlier recorded in the joint paper by Aage Bohr, Ben Mottelson, and David Pines (Bohr et al. 1958). The wide impact of the collective description was also acknowledged by the Japanese physicist H. Nakano, who nominated both Bohm and Pines for the Physics Nobel Prize in 1958.[11] According to Pines (1987, p. 78), from his work with Bohm emerged what was called at first the plasma approximation and later the "random phase approximation." The wide use of this method is demonstrated by the fact that 7150 papers appear on the Web of Science with it as a topic. These works using the collective variables approach are therefore part of Bohm's lasting contribution to physics.

Let us now consider the case of the topic "Bohm diffusion," which is a reference to the equation Bohm had suggested in his early work on plasma, still at Berkeley. The equation is the following: $D_\perp \approx \frac{10^5}{16H}\left(\frac{kTe}{e}\right)$. In the early 1960s, following the ongoing interest in harnessing nuclear fusion for the purposes of obtaining cleaner and more powerful sources of nuclear energy, physicists went back to use Bohm's result which had been experimentally confirmed thanks to technical improvements in this field of physics. The term has been cited, as a topic, so far, by 345 papers. It is thus part of the influential results Bohm obtained with plasma at the end of WWII at Berkeley, which were later resumed at Princeton through the collaboration with Gross and Pines. However, which of Bohm's paper is being cited when Bohm's diffusion is cited? In hindsight we can say that it is the book chapter "The Use of Probes for Plasma Exploration in Strong Magnetic Fields," co-authored by Bohm, Eric Burhop, and Harrie Massey. This chapter was published in 1949 in the book *The Characteristics of Electrical Discharges in Magnetic Fields*, which was edited by A. Guthrie and R. K. Wakerling, (eds), reporting the work developed in the war effort.[12] However, this poses a problem for studies based on scientometry as the main base of data indexes articles published in journals and not book chapters. Furthermore, the

---

[10]Ben Mottelson, The Nobel Lecture, available at: https://www.nobelprize.org/nobel_prizes/physics/laureates/1975/mottelson-lecture.html.

[11]See: https://www.nobelprize.org/nomination/redirector/?redir=archive/. Accessed on 22 Jan 2019.

[12]Bohm et al. (1949), Guthrie and Wakerling (1949).

first references to Bohm diffusion in papers did not make any reference to a paper by ... Bohm. From 1963 to 1990 Bohm diffusion appeared as a topic being cited, many times in the title of the papers, by 13 papers, without any reference to a paper by Bohm himself. It was only in 1990 that H. R. Kaufman, explaining why Bohm diffusion was such a good approximation, cited the original paper. Bohm had presented his result as a "rough estimate." Kaufman concluded his analysis, about 50 years later, saying: "It appears likely that the explanation given above [by Kaufman] for Bohm diffusion is the sort of 'rough estimate' made by Bohm. With the passage of time, Bohm diffusion appears more widely applicable and the justification for this sort of explanation appears more sound."[13]

We may learn two lessons from this. First, of course, Bohm's equation for diffusion in plasma, and his 1949 paper, should be considered among the most influential of Bohm's results and papers. Second, the use of scientometry requires careful and critical examination as even in a field as physics it may hide important information on the production of science.

8.5 Reinterpretation of Quantum Mechanics

Now we examine the three papers on the reinterpretation of quantum physics, also in chronological order of publication. The first two were published together and the number of citations of the first one is the best index of the influence of these two papers. This is because many people may have cited it thinking of the causal interpretation in general. The third paper, co-authored with Vigier, is a development of this interpretation. Thus let us consider the evolution of citations of the first paper, plotted in Fig. 8.7. After an initial number of citations, most criticizing the suggested interpretation, the paper went ignored for almost ten years. It was revived from the 1970s on thanks to the growing interest in the subject of hidden variables in quantum

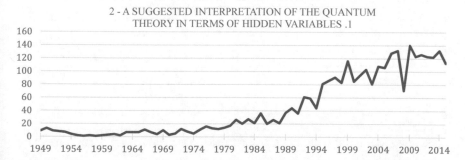

Fig. 8.7 Graph of the number of citations over the years for the paper "A suggested interpretation of the quantum theory in terms of hidden variables—1," by Bohm (1952a). Data from Web of Science—Core Collection, ranging from 1945 to 2018, accessed on 07 December 2018

---

[13] Kaufman (1990).

mechanics following the publication of Bell's theorem, in 1965, and the appearance of its early experimental tests, which contrasted quantum mechanics with local realism, as we saw in Chaps. 4 and 5.

Interest in Bohm's proposal in the 1980s continued to increase due to other factors. On the one hand, it may have been a side effect of interest in Alain Aspect's 1981–1982 experiments. These experiments reinforced physicists' trust in quantum mechanics predictions, against the possibility of local realist theories. However, the debates triggered by Aspect's experiments recalled Bohm's interpretation. Indeed it had survived these tests as it was as nonlocal as quantum mechanics was. On the other hand, this interest was reinforced when Chris Philippidis, Chris Dewdney, and Basil Hiley published the first diagrams of the causal interpretation paths as well as those of Bohm's quantum potential. These diagrams were produced by computers and had no new physical equations. However, pictures are always powerful tools for communication, in particular as a way to view weird or unknown physical phenomena, as emphasized by historian Peter Galison in his book on the material culture of 20th century microphysics and by historian David Kaiser in his study of the dispersion of Feynman's diagrams. Thus Bohm's diagrams provided a conceptual advantage to Bohm's hidden variable interpretation. These diagrams are now an iconic representation of Bohm's proposal and many argue that they brought more intelligibility to quantum physics. Finally from the 1990s to nowadays, citations of this paper have continued to increase influenced by the appearance of Bohmian mechanics, a slightly modified presentation of Bohm's early ideas advanced by Sheldon Goldstein and by the wide interest in foundations of quantum mechanics following the blossoming of quantum information in the mid-1990s.[14]

This pattern for Bohm's 1952 hidden variables interpretation, mostly from 1970 on, is reproduced when we plot the number of citations over time for the other papers (Papers 3 and 8) about the reinterpretation of quantum mechanics, see Figs. 8.8 and 8.9.

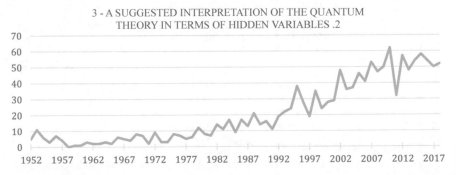

Fig. 8.8 Graph of the number of citations over the years for the paper "A suggested interpretation of the quantum theory in terms of hidden variables—2," by Bohm (1952b). Data from Web of Science—Core Collection, ranging from 1945 to 2018, accessed on 07 December 2018

---

[14](Philippidis et al. 1979; Galison 1997; Kaiser 2005).

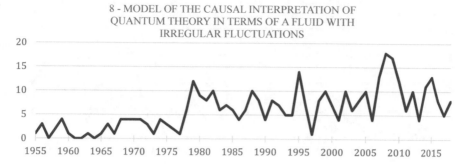

Fig. 8.9 Graph of the number of citations over the years for the paper "Model of the causal interpretation of quantum theory in terms of a fluid with irregular fluctuations," by Bohm and Vigier (1954). Data from Web of Science—Core Collection, ranging from 1945 to 2018, accessed on 07 December 2018

8.6   Quantum Effects and Experiments

We finally move to the papers which emerged from the collaboration between Bohm and his doctoral student Yakir Aharonov. Their first paper addressed the question: is there an experiment whose results may be used to portray the EPR argument? They analyzed an experiment conducted by the Sino-American physicist Madame Chien Shiung Wu for a different purpose and showed that the EPR argument ran counter to quantum mechanics predictions. Bohm and Aharonov demonstrated that empirical data obtained from Wu's experiment corroborated quantum mechanics and violated the EPR argument. In the first years after its publication, this paper did not attract much attention but later, after interest was awakened by Bell's theorem experiments, the paper received more citations. It is mostly used as an example of the first attempt to check the EPR argument against quantum mechanics predictions.[15] Thus the readers will note that the pattern of the rising number of citations of this paper, plotted in Fig. 8.10, has some similarity with the patterns of the papers on the reinterpretation of quantum mechanics which we discussed in the previous section

Two years later, in 1959, Bohm's most cited paper so far was published. It presented what is known as the Aharonov-Bohm effect. Aharonov was its first author reflecting his share of work in the solution of the problem, the subject of his doctoral thesis. Aharonov and Bohm challenged the idea, current in classical electromagnetism, according to which the vector potential has no physical significance. They showed that an electron beam travelling in a region free from any electrical field but near a solenoid where a magnetic field is confined may still undergo influence due to the variation in the confined magnetic field. This paper started an industry devoted to discussing its theoretical and experimental implications. Theoretical issues concerned whether this effect was strictly quantum or not, if it was topological in nature, and if it was an illustration of quantum non-locality. Researchers produced

---

[15]Wu and Shaknov (1950). On this experiment, see (Maia Filho and Silva 2019).

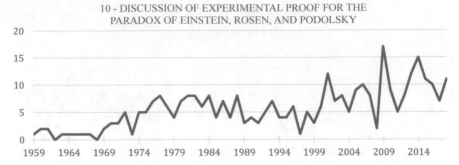

Fig. 8.10 Graph of the number of citations over the years for the paper "Discussion of experimental proof for the paradox of Einstein, Rosen, and Podolsky," by Bohm and Aharonov (1957). Data from Web of Science—Core Collection, ranging from 1945 to 2018, accessed on 07 December 2018

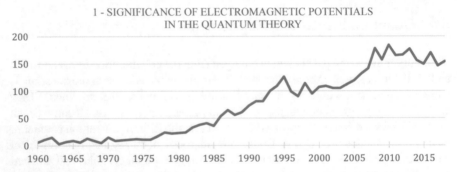

Fig. 8.11 Graph of the number of citations over the years for the paper "Significance of electromagnetic potentials in the quantum theory," by Aharonov and Bohm (1959). Data from Web of Science—Core Collection, ranging from 1945 to 2018, accessed on 07 December 2018

a string of experiments, mainly led by Peshkin and Tonomura, whose results convinced physicists of the real existence of such an effect. It is this whole industry which is responsible for the huge and ever increasing number of citations of this paper, see Fig. 8.11. The paper was the basis for the nominations both of Bohm and Aharonov for several scientific accolades, which included the Wolf Prize in Physics for Aharonov in 1998.

8.7   Qualitative Comments

All this scientometric data may now be tempered with evidence from the views of some of Bohm's contemporaries on his work. Views on Bohm's achievement were corroborated by some of his greatest collaborators. In 1989, when Basil Hiley was asked by his head of department to send some background information in order to support Bohm's nomination for the Nobel Prize for quantum non-locality, he chose precisely the following three fields to single out Bohm's achievements. According to Hiley's letter, "the first major contribution made by Bohm was in the area of plasma" and "the second area in which Bohm has contributed significantly is our understanding of quantum mechanics." In this latter area Hiley cited both the paper on the Aharonov-Bohm effect and the papers on the reinterpretation of quantum mechanics.[16]

Let me concentrate now on the cases of his papers concerning the reinterpretation of quantum mechanics. Well known are John Bell's reminiscences about the influences shaping his deep interest in the foundations of quantum mechanics. In 1982, he wrote "Bohm's 1952 papers on quantum mechanics were for me a revelation," and "In 1952 I saw the impossible done." Thus Bohm's causal interpretation is part of the history of Bell's theorem. Earlier, Bohm had been compared to Kepler. Indeed, in 1958, in a fierce exchange of correspondence with Léon Rosenfeld about Bohm's works and their worth, Lancelot Whyte wrote: "Naturally you are fully aware … that valuable results may spring from mistaken motives and reasoning. Kepler is a good example. But this awareness is not evident in your review."[17] It is not bad for a physicist to be compared to Kepler. Jumping in time, after Bohm's passing, evidence of the influence of his works may be seen in the very fact that there is an intellectual diaspora of physicists, discussed in Chap. 7, who were influenced by Bohm's causal interpretation. Just to illustrate, physicists such as Sheldon Goldstein, Basil Hiley, N. Pinto-Neto, and Antony Valentini disagree about the meaning and reach of Bohm's original works on the reinterpretation of quantum mechanics. Only major thinkers may have this kind of diaspora as their offspring.

Thus, recognition of Bohm's contribution to the enlightenment of the foundations of quantum mechanics closely follows the line of citations of his 1952 papers. Both reflect the changing views of physicists about the role and value of research on these topics. It is against the backdrop of these changing views that we may understand Melba Phillips' comment: "It is too bad, very sad indeed, that he did not live to see how his reputation has shot up recently. His interpretation of quantum mechanics is becoming respected not only by philosophers of science but also by 'straight' physicists." Phillips, who was his close friend for most of his life, wrote this in 1994, and Bohm had passed away in 1992.[18] This was the time when the field of quantum information began to take off, drawing from the resources exploited in the research on the foundations of quantum mechanics.

While Bohm's work on the reinterpretation of quantum mechanics was responsible for most of his influence on physics, one should not underestimate the influence of the other two fields where Bohm excelled, Collective Variables and Quantum Effects. As for the work on Collective Variables, we have seen the acknowledgment of Ben

---

[16]Basil Hiley to Sessler, 9 January 1989. Folder A172, David Bohm Papers, Birkbeck College, London.

[17]All these citations are in Freire Junior (2015).

[18]Melba Phillips to David Peat, 17 Oct 1994, folder A22, Bohm Papers. Birkbeck College, London. On the relationship between Bohm and Phillips, see their letters in Talbot (2017).

Mottelson on the occasion of the Nobel Prize, and Nakano's nominating Bohm and Pines for the Nobel Prize. On the work on Quantum Effects, the number of citations of the paper by Aharonov and Bohm is telling of its influence among fellow physicists.

8.8   Concluding Remarks

Let me come back to Bohm's legacy to summarize my conclusions from this analysis:

- Bohm left lasting and influential contributions in all three domains we have considered here: Collective Variables, Quantum Effects, and Interpretations of Quantum Mechanics. These contributions rank him as one of the most important theoretical physicists in the 20th century.
- His most influential contributions were published from 1949 to 1959 and there was a decline in the production of influential papers from 1959 on.
- The wholeness program, which was developed from the early 1970s till 1992 and was responsible for Bohm's wide intellectual influence has not yet borne similar fruit among scientists.

To conclude, if I were asked to present Bohm's legacy to physics in a nutshell, I would say he was a physicist who made many and lasting contributions; no doubt one of the greatest 20th century physicists. However, as for his legacy to quantum physics, rather than for one specific and lasting contribution, I think he should be acknowledged for his pointing out the relevance of the research on the foundations of this theory, which includes both his works on the Interpretation of Quantum Mechanics and Quantum Effects.

References

Aharonov, Y., Bohm, D.: Significance of electromagnetic potentials in the quantum theory. Phys. Rev. **115**(3), 485–491 (1959)

Aharonov, Y., Bohm, D.: Time in quantum theory and uncertainty relation for time and energy. Phys. Rev. **122**(5), 1649–1658 (1961a)

Aharonov, Y., Bohm, D.: Further considerations on electromagnetic potentials in quantum theory. Phys. Rev. **123**(4), 1511–1524 (1961b)

Bohm, D.: A suggested interpretation of the quantum theory in terms of hidden variables-I. Phys. Rev. **85**(2), 166–179 (1952a)

Bohm, D.: A suggested interpretation of the quantum theory in terms of hidden variables-II. Phys. Rev. **85**(2), 180–193 (1952b)

Bohm, D.: Proof that probability density approaches Ψ^2 in causal interpretation of the quantum theory. Phys. Rev. **89**(2), 458–466 (1953)

Bohm, D.: Wholeness and the Implicate Order. Routledge & Kegan Paul, London (1980)

Bohm, D., Aharonov, Y.: Discussion of experimental proof for the Paradox of Einstein, Rosen, and Podolsky. Phys. Rev. **108**(4), 1070–1076 (1957)

Bohm, D., Bub, J.: A proposed solution of measurement problem in quantum mechanics by a hidden variable theory. Rev. Mod. Phys. **38**(3), 453–469 (1966)

Bohm, D., Burhop, E.H.S., Massey, H.S.W.: The use of probes for plasma exploration in strong magnetic fields. In: Guthrie, A., Wakerling, R.K. (eds.) The Characteristics of Electrical Discharges in Magnetic Fields, pp. 13–76. McGraw-Hill, New York (1949)

Bohm, D., Gross, E.: Theory of plasma oscillations. A. Origin of medium-like behavior. Phys. Rev. **75**(12), 1851–1864 (1949a)

Bohm, D., Gross, E.: Theory of plasma oscillations. B. Excitation and damping of oscillations. Phys. Rev. **75**(12), 1864–1876 (1949b)

Bohm, D., Hiley, B.J.: The Undivided Universe: An Ontological Interpretation of Quantum Theory. Routledge, London (1993)

Bohm, D., Pines, D.: A collective description of electron interactions. 1. Magnetic Interactions. Phys. Rev. **82**(5), 625–634 (1951)

Bohm, D., Pines, D.: A collective description of electron interactions. A collective description of electron interactions. Phys. Rev. **92**(3), 609–625 (1953)

Bohm, D., Vigier, J.P.: Model of the causal interpretation of quantum theory in terms of a fluid with irregular fluctuations. Phys. Rev. **96**(1), 208–216 (1954)

Bohm, D., Hiley, B.J., Kaloyerou, P.: An ontological basis for the quantum-theory. Phys. Rep. **144**(6), 321–375 (1987)

Bohm, D., Weinstein, M.: The self-oscillations of a charged particle. Phys. Rev. **74**(12), 1789–1798 (1948)

Bohr, A., Mottelson, B.R., Pines, D.: Possible analogy between the excitation spectra of nuclei and those of the superconducting metallic state. Phys. Rev. **110**(4), 936–938 (1958)

Freire Junior, O.: The Quantum Dissidents—Rebuilding the Foundations of Quantum Mechanics 1950–1990. Springer, Berlin (2015)

Freitas, F.H.A., Freire Junior, O.: Sobre o uso da Web of Science como fonte para a história da ciência. Revista da SBHC **1**(2), 129–147 (2003)

Galison, P.: Image and Logic—A Material Culture of Microphysics. Chicago University Press, Chicago (1997)

Guthrie, A., Wakerling, R.K. (eds.): The Characteristics of Electrical Discharges in Magnetic Fields. McGraw-Hill, New York (1949)

Hicks, D., Wouters, P., Waltman, L., de Rijcke, S., Rafols, I.: Bibliometrics: the Leiden Manifesto for research metrics. Nature **520**(7548), 429–531 (2015)

Kaiser, D.: Drawing theories apart: the dispersion of Feynman diagrams in postwar physics. Chicago University Press, Chicago (2005)

Kaufman, H.R.: Explanation of Bohm diffusion. J. Vac. Sci. Technol. B **8**(1), 107–108 (1990)

Kojevnikov, A.: David Bohm and collective movement. Hist. Stud. Phys. Biol. Sci. **33**, 161–192 (2002)

Kragh, H.: An Introduction to the Historiography of Science. Cambridge University Press, Cambridge (1987)

Maia Filho, A., Silva, I.: O experimento WS de 1950 e as suas implicações para a segunda revolução da mecânica quântica, Revista Brasileira de Ensino de Física **41**(2), e2018018 (2019)

Patience, G.S., Patience, C.A., Blais, B., Bertrand, F.: Citation analysis of scientific categories. Heliyon **3**(5), e00300 (2017)

Philippidis, C., Dewdney, C., Hiley, B.J.: Quantum interference and the quantum potential. Nuovo Cimento della Societa Italiana di Fisica B—Gen. Phys. Relat. Astron. Math. Phys. Methods **52**(1), 15–28 (1979)

Pines, D.: A collective description of electron interactions: IV. Electron interaction in metals. Phys. Rev. **92**, 625–636 (1953)

Pines, D.: The collective description of particle interactions: from plasmas to the Helium liquids. In: Hiley, B.J., Peat, F.D. (eds.) Quantum Implications: Essays in Honour of David Bohm, pp. 66–84. Routledge & Kegan Paul, London (1987)

Pines, D., Bohm, D.: A collective description of electron interactions. A collective description of electron interactions. Phys. Rev. **85**(2), 338–353 (1952)

Podlubny, I.: Comparison of scientific impact expressed by the number of citations in different fields of science. Scientometrics **64**(1), 95–99 (2005)

Redner, S.: Citation statistics from 110 years of physical review. Phys. Today, pp. 49–54 (2005)

Talbot, C. (ed.): David Bohm: Causality and Chance, Letters to Three Women. Springer, Berlin (2017)

Wu, C.S., Shaknov, I.: The angular correlation of scattered annihilation radiation. Phys. Rev. **77**(1), 136 (1950)

Author Index

A

Aharonov, Y., 3, 103–105, 107, 110–112, 120, 124, 134, 182, 183, 186, 194, 205, 210, 217, 226, 227, 229, 235–238

Albert, D., 183, 217

Amazing Stories, 19, 20, 22

American Journal of Physics, 51

American Physical Society, 39, 40, 43, 139, 163

Anderson, H. L., 88

Arshansky, R. I., 185

Aspect, A., 13, 15, 156, 158, 159, 164, 169, 179–181, 186, 190, 234

Atkins, P. W., 188

Atmanspacher, H., 210, 217, 218

B

Badino, M., 11

Bardeen, J., 53, 54, 96, 124, 126

Barros, A. de, 101, 211, 212

Barut, A.O., 185

Bastin, E. W., 135, 151

Becker, A., 16, 25, 61, 78

Beck, G., 14, 81, 86, 92

Belinfante, F., 171

Beller, M., 150

Bell, J. S., 2, 4, 5, 7, 12, 13, 49, 66, 67, 119, 140, 141, 156–162, 164, 179–183, 190, 194, 206, 207, 214, 237

Berkeley Radiation Laboratory, 35, 55, 61

Besson, V., 65, 70, 73, 82, 84, 89–93, 99, 108, 134, 212

Biederman, C., 14, 122, 123, 131, 153, 209

Bird, K., 34–37

Birge, R. T., 39, 40

Birkbeck College, 12, 14, 36, 70, 77, 107, 112, 113, 119, 125, 134, 142, 154, 155, 158, 161, 179, 186, 190, 200, 216, 217, 229, 237

Blokhintsev, D. I., 68, 69, 99

Bohm, D., 232

Bohm, F., 20

Bohm, S., 20

Bohr, A., 5, 53, 87, 95, 96, 124, 126, 133, 145, 146, 232

Bohr, N., 1, 2, 9, 14, 16, 20, 21, 33, 39, 48–50, 69, 70, 78, 84, 90, 95, 96, 103, 113, 125, 127, 128, 132, 133, 135, 136, 141, 145–147, 152, 155, 170, 171, 175, 177, 181, 183, 193, 213

Born, M., 66, 71, 84, 85, 87, 94, 97

Bourdieu, P., 8, 13, 164, 209

Božić, M., 185

Briggs, J., 183, 187

Brockwood Park School, 130

Brown, H., 185

Bub, J., 6, 15, 120, 121, 131, 134–136, 138–143, 147, 149, 150, 152, 156, 162, 175, 177, 181, 185, 215, 227

Bunge, M., 80, 91–93, 108, 223

Burhop, E. H. S., 15, 39–41, 86, 87, 98, 108, 229, 232

C

Callaghan, R. E., 217

Caltech, 5, 19, 27, 29, 31–33, 38

Carmi, G., 103, 110–112, 124, 125, 134

Causality and Chance in Modern Physics, 81, 86, 107, 196

© Springer Nature Switzerland AG 2019
O. Freire Junior, *David Bohm*, Springer Biographies,
https://doi.org/10.1007/978-3-030-22715-9

Subject Index

A

Action at a distance and action by continuity, 218, 219

Active information, 169, 177, 191, 192, 194, 197–199, 210, 216

Afroyim v Rusk, 189

Aharonov-Bohm effect, 3, 5, 107, 111, 112, 134, 186, 194, 205, 210, 235

Algebras
  algebraic approach, 217
  algebraic topology, 135, 152
  Clifford algebras, 183, 215–217
  generalized algebras, 154, 197
  quantum algebras, 155

American atomic project, 1, 4, 35, 59, 230

American dream, 19, 25

B

Beables, 181, 214

Bell's theorem
  Bell's inequalities, 157, 158, 162, 179, 180, 194, 206
  Bell's theorem experiments, 5, 156, 179, 206
  Bell's theorem for three particles, 194

Bohm diffusion
  Bohm's legacy, 5, 224, 229, 238
  Bohm's persona, 224

Bohmian mechanics, 193, 198, 199, 205, 210, 212–215, 217, 226, 234

Bohm's trajectories
  graphs of Bohm's trajectories, 3, 171, 174

Bohm-Vigier illness, 86

C

Calutron, 35, 38, 39

Causal interpretation (of quantum mechanics)
  causality, 65

Chance
  contingency, 122

Citations, 5, 9, 33, 37, 41, 42, 53, 54, 56, 85, 103, 104, 110, 112, 134, 171, 175–177, 186, 187, 223–238

Citizenship (American, Brazilian), 1, 6, 35, 77, 99, 101, 133

Cold War, 1, 6, 11, 35, 47, 55–59, 62, 77, 90, 93, 100, 107, 125, 133, 134, 189, 190, 209

Collapse of the wave packet, 136

Collective variables
  collective coordinates, 42, 52, 124, 126, 134, 209, 226
  collective description, 53, 186, 230–232
  collective movement, 41, 42, 52
  collectivism, 42

Colston (Society, symposium), 109, 110

Communism
  Communist Party, 1, 6, 19, 25, 34–38, 57, 59, 62, 85, 93, 105, 106, 127, 133

Complementarity, 2, 4, 43, 48, 49, 64, 66–70, 72, 84–87, 90, 93, 95, 97, 99, 109, 137, 141, 142, 145, 146, 160, 175, 177, 183, 193

Condensed matter physics, 42, 225

Controversies, 9, 11, 63, 71, 84, 89, 98, 99, 109, 154, 162, 163, 218, 219

Cosmology
  cosmological models, 218, 219

© Springer Nature Switzerland AG 2019
O. Freire Junior, *David Bohm*, Springer Biographies,
https://doi.org/10.1007/978-3-030-22715-9

Printed in the United States
By Bookmasters